Power Transformer
Design Practices

Power Transformer Design Practices

Fang Zhu

Baitun Yang

CRC Press
Taylor & Francis Group
Boca Raton London New York

CRC Press is an imprint of the
Taylor & Francis Group, an **informa** business

First edition published 2021
by CRC Press
6000 Broken Sound Parkway NW, Suite 300, Boca Raton, FL 33487-2742

and by CRC Press
2 Park Square, Milton Park, Abingdon, Oxon, OX14 4RN

© 2021 Taylor & Francis Group, LLC

CRC Press is an imprint of Taylor & Francis Group, LLC

Library of Congress Cataloging-in-Publication Data

Names: Zhu, Fang (Electrical engineer), author. | Yang, Baitun, author.
Title: Power transformer design practices / Fang Zhu, Baitun Yang.
Description: First edition. | Boca Raton, FL : CRC Press/Taylor & Francis Group, LLC, 2021. | Includes bibliographical references and index.
Identifiers: LCCN 2020046765 (print) | LCCN 2020046766 (ebook) | ISBN 9780367418434 (hardback) | ISBN 9780367816865 (ebook)
Subjects: LCSH: Electric transformers. | Electric power distribution--Equipment and supplies.
Classification: LCC TK2551 .Z45 2021 (print) | LCC TK2551 (ebook) | DDC 621.31/4--dc23
LC record available at https://lccn.loc.gov/2020046765
LC ebook record available at https://lccn.loc.gov/2020046766

ISBN: 978-0-367-41843-4 (hbk)
ISBN: 978-0-367-72341-5 (pbk)
ISBN: 978-0-367-81686-5 (ebk)

Typeset in Times
by SPi Global, India

To our parents
Zhu Mingshan, Wang Xiuying, Yang Jiaowen, and Li Aiju
In gratitude for their raising and support.

Contents

Preface

This book presents main topics of practice in power transformer designing such as core, winding, insulation, impedance, losses, cooling, short circuit withstand abilities, sound level and special concerns on auto transformer, and testing. Each of these topics deals with basic theory, design principles, methods and examples, with the intent to help readers on their works. Power transformers operate under a certain circumstance and conditions, so the design and test conducted have to meet certain standards to satisfy the operational request. IEEE and IEC standards are referenced in the book as examples, and the book also presents a guide on how to apply a standard in practice works.

The progress of power transformer engineering is not spectacular but results from continuous small improvements. Better performance, effective application of material are the goals. Performance-wise, losses calculation and reduction methods, cooling calculations and effective designs, measures to make winding strong enough against short circuit forces, sound level calculations and reduction methods and function and size of tertiary winding in auto transformer are presented. The way to effectively use materials meanwhile offering the same performance and reliability are discussed. The purpose, setup and general criteria of tests are present, solutions to issues incurred in test and guide to failure investigation are offered.

This book is a summary of our work in power transformer manufacturing and engineering over many years. During these years, we have gained tremendous knowledge and experience involving a lot of challenges to designing, production processing, research, testing and troubleshooting on a daily basis, and seen the solutions to these challenges whether they are successful or not. The purpose of this book is to present knowledge and experience, and we wish it helps someone improve their engineering knowledge towards perfection.

Fang Zhu

Baitun Yang

Authors

Dr. Fang Zhu received her BSEE and MS degrees in electrical engineering from Xian Jiaotong University in China and her PhD degree in electrical engineering from Strathclyde University, UK. Her major work field is electrical design of power transformers. She worked for Haefely-Trench High Voltage Tech, VA Tech Ferranti-Packard Transformer Ltd., ABB Guelph, Hammond Power Solutions Inc. in Canada, and Pennsylvania Transformer Technology Inc., USA. She is currently engineering manager at R. E. Uptegraff, USA. She is a professional engineer in Ontario, Canada.

Dr. Baitun Yang received his BSEE, MS and PhD degrees in electrical engineering from Xian Jiaotong University in China. His work field is in transformer testing, failure investigation, insulation structure and transient analysis. He worked for Northern Transformer, VA Tech Ferranti-Packard Transformer Ltd., Hammond Power Solutions Inc. in Canada, and Pennsylvania Transformer Technology Inc. in the USA. He is currently senior engineer at R. E. Uptegraff, USA. He is a member of the IEEE PES transformers committee.

1 Introduction

Power transformers are an important component in electrical power transmission networks. Several types of power transformers exist, such as core form and shell form regarding the relation between winding and core. Under the core form specialty, there are three-leg core, five-leg core and single-leg core transformers. This book deals with three-leg core form transformers, which are widely used.

1.1 BASIC THEORY

The basic function of the power transformer is changing the system voltage from one level to another through electromagnetic interaction, so the Law of Electromagnetic Induction and some parameters which are important in transformer design are introduced here.

1.1.1 VOLTAGE AND CURRENT OF WINDINGS

A winding having N turns is placed in a magnetic field, assuming each of its turns is linked with the same magnetic flux, ϕ, which varies with time, t. By the Law of Electromagnetic Induction, the voltage induced in the winding, $v(t)$, is

$$v(t) = -N(d\phi / dt) \tag{1.1}$$

If ϕ is sinusoidally varying flux with amplitude ϕ_m and frequency f

$$\phi = \phi_m \cos(2\pi ft) \tag{1.2}$$

Equation (1.1) is changed to

$$v(t) = 2\pi fN\phi_m \sin(2\pi ft) = \sqrt{2}V\sin(2\pi ft) \tag{1.3}$$

Where V is the effective value of the induced voltage, or voltage as called therein after. It is deducted from Equation (1.3).

$$(V / N) = \sqrt{2}\pi f\phi_m = 4.44 f B_m A_{Fe}; \quad B_{max} = \frac{(V / N)}{4.44 f A_{Fe}} \tag{1.4}$$

Where B_{max} is maximum flux density in the core, Tesla; A_{Fe} is the net core cross-sectional area the flux going through, m², $A_{Fe} = F_{stack} \times A$, where F_{stack} is stacking factor considering lamination coat thickness, it is in the range of 0.96~0.97, A is the core cross-sectional area, m². f is frequency, Hz. (V/N) is called volts per turn. Equation (1.4) gives the relation between magnetic performance and electric performance of a transformer. In a practical transformer, the primary and secondary windings are

1

wound around a steel core in which the mutual flux goes through, and the great magnetic properties of core steel greatly increases the mutual flux density compared to air core. As results, (V/N) can be increased to handle high voltage or large power rating units, or in another way, with the same (V/N) the core cross-sectional area can be reduced dramatically, resulting in size reduction of the core, the windings, and overall size of the unit.

Understanding a real transformer starts usually from understanding an ideal transformer. An ideal two-winding transformer is a transformer which meets the following assumptions. First, both primary and secondary windings are linked only to mutual flux in the core; there is no flux field outside the core. Secondly, the core doesn't need exciting current to magnetize it and doesn't consume power or energy. Third, neither of the windings consumes energy, meaning that the windings have no resistances. In an ideal transformer, input power is equal to output power, $V_1 I_1 = V_2 I_2$, the voltages and the currents of winding 1 and winding 2, shown in Figure 1.1, have relations which can be deducted from Equation (1.4).

$$\frac{V_2}{V_1} = \frac{N_2}{N_1}; \frac{I_1}{I_2} = \frac{N_2}{N_1} \tag{1.5}$$

In a real transformer, first, there is a flux field outside the core; it is generated by the currents in the windings and exists in and between the windings. Called leakage flux, it produces eddy losses in the winding conductor as well as in other metallic parts such as the tank wall and core clamping structure. Second, exciting the core consumes energy and current because the core steel is not loss-free. Third, a current going through winding produces I^2R loss due to existence of resistance of the winding material. Further, the core steel has magnetic nonlinearity caused by the effect of magnetic hysteresis; this means that the flux in the core is not linearly proportional to the exciting current, as shown in Figure 1.2. Power transformers are normally

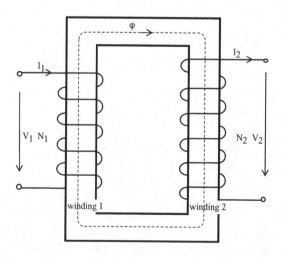

FIGURE 1.1 The transformation of voltages and currents.

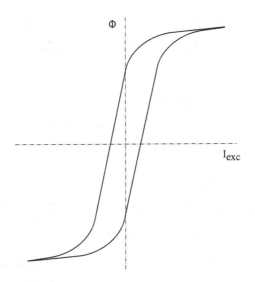

FIGURE 1.2   Flux and its exciting current.

operated with sinusoidal primary voltage, the flux density in the core should be sinu-soidal if sinusoidal secondary voltage is desired, then the exciting current, due to nonlinearity of the core steel shown in Figure 1.2, is non-sinusoidal.

An equivalent circuit of one phase of a real transformer presenting all of these characteristics is shown in Figure 1.3a. Unlike the flux in steel core which is not linearly proportional to its exciting current, the leakage flux has a considerable por-tion of its path in oil or air, which makes the leakage flux be linearly proportional to the current producing it. This means the leakage impedance has a constant value. In a real transformer, the exciting current is so small compared to load current of the primary side that the exciting current is often neglected entirely, such that the real transformer has the properties of linear circuit because the winding resistance and leakage impedance are constants. The exciting current is also assumed to be an equivalent sine wave, except those studies which directly relate to exciting current such as harmonics analysis. Then the equivalent circuit can be simplified as shown in Figure 1.3b and c and is analyzed by simple vector methods.

In the real transformer, turn ratio (N_1/N_2) and induced voltage (electromotive force) ratio $(e_1/e_2$ in Figure 1.3a) are same. Under no-load conditions, the terminal voltage ratio (V_1/V_2) is nearly equal to turn ratio because the exciting current I_{exc} is so small that the voltage drop on leakage impedance of R_1 and X_1 is negligible. Under load conditions, the voltage drops on leakage impedance may be noticeable, and the difference between turn ratio (N_1/N_2) and terminal voltage ratio (V_1/V_2) may be observable, too. Secondly, since magnetizing the core needs current I_{exc}, theoretically speaking, the current ratio (I_1/I_2) is not equal to inverse turn ratio (N_2/N_1). However, due to its very small value, I_{exc} is often neglected entirely. The current ratio relation with turn ratio in Equation (1.5) is still used in power transformer designs satisfactorily.

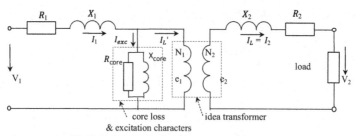

(a) Equivalent circuit of real iron-core transformer

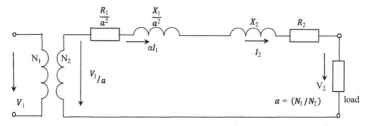

(b) Equivalent circuit referred to secondary side

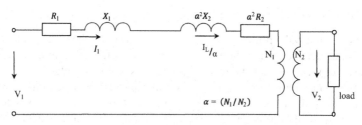

(c) Equivalent circuit referred to primary side

FIGURE 1.3   Equivalent circuits of a real transformer.

Transformer resistance and reactance are often expressed in percentages as.

$$R\% = \frac{V_R}{V} = \frac{LL}{VI}; X\% = \frac{V_X}{V} = \frac{Q}{VI} \tag{1.6}$$

Where $R\%$ and $X\%$ are the percentages of resistance and reactance respectively, V_R and V_X are resistance and reactance voltage drops respectively, V is rated phase voltage, I is rated phase current, LL is load loss, Q is reactive power.

Notice that the transformer doesn't transfer DC current and voltage; this means that equations previously listed apply only to AC current and voltage. Therefore, it may be said that a transformer is a device which transfers AC current and AC voltage. The transformer may be used to prevent DC current and voltage transferred from one circuit into another circuit, in the meantime maintaining alternating current continuity between the circuits.

1.1.2 LOSSES

When a transformer is energized, its magnetized core consumes energy because the core steel material is not loss-free. The consumed energy in form of heat is called no-load loss. The no-load loss exists as long as the unit is energized regardless of whether it carries a load or not. To reduce the no-load loss, cold-rolled grain-oriented core steel is used in present. Its loss per unit weight is greatly reduced compared to hot-rolled, non-oriented core steel.

When a transformer carries a load, the loading currents in both primary and secondary windings generate I^2R losses, the leakage flux produced by the loading currents hitting winding conductors and metallic structure parts such as tank wall and core frame generates eddy current loss, sum of these losses is called load loss and is in form of heat. The major portion of the load loss is I^2R loss, sometimes called DC loss. The commonly used calculation equation of I^2R loss per kilogram copper material is

$$I^2R = \frac{J^2 \cdot \rho \times 10^9}{d_{Cu}} \left(W / kg \right) \tag{1.7}$$

Where ρ is copper resistivity, 1.724×10^{-8} $\Omega \cdot m$ at 20°C, d_{Cu} is copper density, 8.89 kg/dm³, J is current density, A/mm². At 75°C, copper resistivity is 2.097×10^{-8} $\Omega \cdot m$, I^2R loss per kilogram copper is then

$$I^2R = 2.36 \cdot J^2 \left(W / kg \right) \tag{1.8}$$

Besides I^2R loss, the rest of load loss consists of eddy current losses in winding, and in structural parts, this potion of load loss is related to the leakage flux.

1.1.3 MAXIMUM LEAKAGE FLUX DENSITY

Different from the flux in core, the leakage flux is generated by winding currents; it exists outside the core and in the space inside the tank of transformer. Its distribution in the middle of a two-winding unit is shown in Figure 1.4. The leakage flux density reaches its maximum, $B_{leakage\,max}$, in the gap between windings. Assuming the ampere-turns of both windings are evenly distributed along the winding height, and the effect of the winding ends on the leakage flux is ignored, $B_{leakage\,max}$ deducted from Maxwell equation is as

$$B_{leakage\,max} = \sqrt{2} \cdot \mu_0 \cdot \frac{I \cdot N}{H_{wdg}} \left(Tesla \right) \tag{1.9}$$

where $\mu_0 = 4\pi \times 10^{-7}$ H/m, vacuum permeability, $(I \cdot N)$ is the ampere-turn of either winding 1 or winding 2, I is the rms value of the winding current, H_{wdg} is average magnetic height of windings, m. Another way to express the maximum leakage flux density is

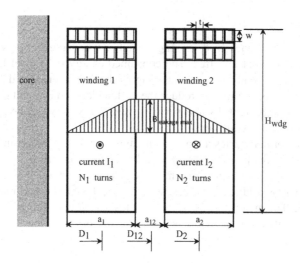

FIGURE 1.4 Leakage flux distribution between two windings.

$$B_{leakage\,max} = \sqrt{2} \cdot \mu_0 \cdot \frac{\left(I/s\right)s \cdot N_{sec} \cdot N_{T/sec}}{H_{wdg}}$$

$$= \sqrt{2} \cdot \mu_0 \cdot \frac{J \cdot \left(N_{sec} \cdot w\right) \cdot \left(N_{T/sec} \cdot t\right)}{H_{wdg}} \tag{1.10}$$

$$\approx \sqrt{2} \cdot \mu_0 \cdot J \cdot RB \cdot S_f \times 10^3 \,(\text{Tesla})$$

where $N_{sec} \cdot N_{T/sec} = N$, N_{sec} is number of sections of winding, $N_{T/sec}$ is number of turns per section, RB is winding radial build, $RB \approx N_{T/sec} \cdot t$, where t is conductor radial thickness of one turn, mm; J is current density, A/mm²; $S_f = (N_{sec} \cdot w)/H_{wdg}$, related to winding space factor, where w is conductor axial height of one turn. Here, the assumptions are applied that cable thickness and height are approximately equal to its conductor thickness and height, and that total thickness of radial spacers in winding is negligible compared to total conductor thickness. The leakage flux has positive and negative effects on transformer performance. High leakage flux makes high leakage impedance, commonly called impedance; high impedance helps to reduce short-circuit forces on the windings by reducing short-circuit currents. However, high impedance makes secondary voltage drop even greater, and it causes higher eddy current loss in winding, tank and core clamping structure, resulting in high load loss and possibly a local overheating problem.

1.1.4 IMPEDANCE

The impedance discussed here is positive or negative sequence short-circuit impedance. The leakage flux in the winding presents a reactance behavior, while the load loss presents a resistance behavior. The impedance of a transformer is then

$$Z\% = \sqrt{\left(R\%\right)^2 + \left(X\%\right)^2} \tag{1.11}$$

Where $R\%$ is resistance in percentage and $X\%$ is reactance in percentage. The impedance is an important parameter of a transformer. The impedance exists between pairs of windings and has physical meaning. Sometimes in order to simplify calculations such as single phase line-to-ground short-circuit currents calculation, a three-winding unit is represented as a star equivalent impedance network, and the impedance of each winding is calculated. This is just a mathematical treatment; the impedance of a single winding has no physical meaning. The reactance can be also expressed as.

$$X\% = 0.124 \times \frac{\left(KVA / \phi\right)}{H_{wdg} \cdot \left(V\!/\!N\right)^2} \cdot \frac{f}{50} \cdot \Sigma Da \cdot R_g \tag{1.12}$$

Where (KVA/ϕ) is KVA rating per phase, (V/N) is voltage per turn. H_{wdg} is average magnetic height of windings, mm, f is frequency, Hz, $\Sigma Da = (D_1 \times a_1)/3 + (D_{12} \times a_{12}) + (D_2 \times a_2)/3$, where D_i and a_i, mm, are dimension shown in Figure 1.4, R_g is Rogowski correction factor considering effect of winding ends on leakage flux. The deduction will be carried out in Chapter 5. In general, a reduction in reactance is associated with increased mass of core steel and increased no-load loss and reduced mass of copper and load loss. Conversely, an increase in reactance is associated with reduced no-load loss and increased load loss.

Example 1.1

A three-phase two-winding transformer has 280 MVA rated power. HV has 230 kV line-to-line voltage wye connection, LV has a delta connection. The no-load loss is NLL = 121 kW, the load loss is LL = 933 kW, the impedance is $Z\% = 22.5\%$, the exciting current is $I_{exc} = 0.13\%$. All of the calculations conducted below are referred to 230 kV side.

Method 1

HV rated winding current, I_1, is
$$I_1 = \left(280 \times 10^3\right) / \left(\sqrt{3} \times 230\right) = 702.9\,(A).$$
The effective resistance per phase in ohm is
$$R = LL / \left(3 \times I_1^2\right) = \left(933 \times 10^3\right) / \left(3 \times 702.9^2\right) = 0.629\,(\Omega)$$
This resistance in percentage is
$$R\% = V_R / V_1 = \left(0.629 \times 702.9\right) / \left(230 \times 10^3 / \sqrt{3}\right) = 0.33\%$$
Where V_1 is HV phase voltage. Another way to calculate this resistance is
$$R\% = LL / S_r = 933 \times 10^3 / 280 \times 10^6 = 0.33\%$$
Where S_r is the rated power. The impedance per phase in Ohm is
$$Z = \left(Z\%\right) \times \frac{V_1}{I_1} = 22.5\% \times \frac{\left(230 \times 10^3 / \sqrt{3}\right)}{702.9} = 42.5\,(\Omega)$$
The reactance per phase is
$$X = \sqrt{Z^2 - R^2} = \sqrt{42.5^2 - 0.629^2} \approx 42.5\,(\Omega)$$

R_{core}, Z_{core} and X_{core}, which present the core resistance, impedance and reactance respectively, are

$$R_{core} = \frac{V_1^2}{(NLL/3)} = \left(\frac{230 \times 10^3}{\sqrt{3}}\right)^2 \times \frac{3}{121 \times 10^3} = 437190\,(\Omega/\text{phase})$$

$$Z_{core} = \frac{V_1}{I_{exc}} = \left(\frac{230 \times 10^3}{\sqrt{3}}\right) \times \frac{1}{0.13\% \times 702.9} = 145321\,(\Omega/\text{phase})$$

$$X_{core} = \frac{1}{\sqrt{(1/Z_{core})^2 - (1/R_{core})^2}} = 154082\,(\Omega/\text{phase})$$

It should be noted that the magnetizing reactance, X_{core}, is not constant but is voltage independent due to B-H curve nonlinearity of core material as shown in Figure 1.2. However, when the flux density is below saturation value, X_{core} is treated as a constant in most engineering calculations. When the flux density is higher than the saturation flux density, the flux spills out of the core, and X_{core} is reduced dramatically because the core behaves as air core.

Method 2

The transformer reference impedance, Z_{ref}, is defined as

$$Z_{ref} = (V_{L-L})^2 / S_r$$

Where V_{L-L} is rated line-to-line voltage, V; S_r is rated power, VA. The reference impedance per phase of the unit in this example is

$$Z_{ref} = (230 \times 10^3)^2 / (280 \times 10^6) = 188.93\,(\Omega)$$

The effective resistance per phase is

$$R = (LL/S_r) \cdot Z_{ref} = (933 \times 10^3)/(280 \times 10^6) \times 188.93 = 0.629\,(\Omega)$$

The impedance per phase is Z = 22.5 % × 188.93 = 42.5 (Ω)

1.1.5 Efficiency and Voltage Regulation

When energy is transmitted through a transformer, part of it is consumed by the transformer as no-load loss and load loss. So the output energy is less than energy from input. The efficiency is a measurement of the amount of energy consumed in the transformer and is defined as

$$\eta = \frac{output\ active\ power}{output\ active\ power + (NLL + LL)} = \frac{S_r \cdot \cos\phi}{S_r \cdot \cos\phi + (NLL + LL)} \qquad (1.13)$$

Where η is efficiency, S_r is rated power, $\cos\phi$ is load power factor, NLL is no-load loss, LL is load loss. With different load, the load loss is different, and it makes efficiency change with load, as shown in Figure 1.5. The maximum efficiency occurs when $NLL = LL$, which is hard to achieve even at base rating for power transformer due to LL usually

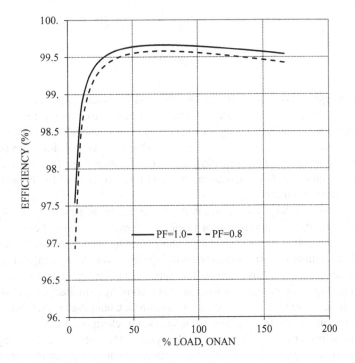

FIGURE 1.5 Relation of efficiency to load.

being much higher than *NLL*. Most units which operate at rated power with loads having an equal or greater than 0.8 power factor have greater than 99% efficiency.

When a transformer is energized but has no load on secondary terminals, the secondary voltage is proportional to primary voltage by turn ratio, as shown in Equation (1.5), neglecting the voltage drop caused by exciting current. When the secondary terminals are connected to a load and load currents flow through the transformer's windings, the secondary voltage has a drop caused by winding resistance and reactance between windings; this drop makes the secondary voltage deviate away from the rated voltage. This drop is called voltage regulation, it is expressed in percentage as follows, with constant primary voltage [1]

$$Voltage\ regulation, \% = \frac{e_2 - V_2}{V_2} \qquad (1.14)$$

Where e_2 is open-circuit secondary voltage, $e_2 = V_1/a$, $a = (N_1/N_2)$, V_2 is secondary voltage under load, $V_2 = e_2 - [(I_2R)^2 + (I_2X)^2]^{0.5}$, where $R = (R_1/a^2 + R_2)$ and $X = (X_1/a^2 + X_2)$; the meanings of these parameters are shown in Figure 1.3. The magnitude of V_2 changes with power flow (load current, I_2) and load power factor ($cos\phi$). In reality, power flows are constantly changing in a network, which makes a constant variation of V_2 away from its rated voltage. Equipment using electrical energy has the best performance at its rated voltage; going away from the rated voltage too much causes the equipment to perform poorly or even not at all, so the output voltage must be kept as constant as possible. In a time period $t_1 - t_2$, the voltage variation, $\Delta V(t)$, can be expressed as

$$\Delta V(t) = (average\,\Delta V) + \sigma(t) \qquad (1.15)$$

Where $(average\,\Delta V)$ is average deviation value and is constant, $\sigma(t)$ is deviation with time and is variable. From Equation (1.15), the voltage variation with time consists of two components: $(average\,\Delta V)$, the average deviation from the rated voltage; and $\sigma(t)$, the fluctuation above and below $(average\,\Delta V)$. The average deviation $(average\,\Delta V)$, can be compensated by adjusting the transformation ratio by means of a de-energized tap changer. The fluctuation, $\sigma(t)$, can be compensated by an on-load tap changer which costs more.

Making output voltage closed to rated voltage can be achieved by adjusting the transformation ratio by tapping off or tapping in turns of winding. Several taps are made in main winding or a separate regulating winding, and between the taps there are a certain number of turns which meet the specified regulating range. These taps are then connected to the tap changer. Two types of tap changer are used, one is the de-energized tap changer which is located in HV winding. Another is the on-load tap changer which is used for either LV or HV windings. Changing taps or changing the number of turns of winding by the de-energized tap changer can be achieved only after the transformer is switched off from the system. Most of de-energized taps have ±5% regulation range in ±2 steps, the tap sections having a tap gap in it are usually located in middle or in the ¼ and ¾ height zones of HV winding where the transient voltage across the tap section could be minimum [2]. When tap sections are in the main winding, two issues need attention regarding the tapping zone, first the voltage reflection from taps which are not in the circuit, this reflection may sometimes result in high voltage in the tap zone under either power frequency or transient voltages at the terminals. Second, when some taps are off-circuited, the sections or a portion of winding connected to these taps is also off-circuited, which means there is no current flowing in this portion of winding. As a result the ampere-turn balance between this winding and adjacent winding get worse, causing higher short-circuit axial forces. When the regulation range is greater than ±5%, the number of tap sections may have to be increased, which makes ampere-turn balance even worse, and the arrangement of taps in main winding become more complicated. These two issues could lead to a separate tap winding, which adds additional cost.

On-load tap changing avoids interruption to the power supply and allows continuous adjustment of the transformation ratio. For application of the on-load tap changer, a separate tap winding is usually needed. The regulation range of the on-load tap changer is wider; for example, ±10~20% in 9 to 35 taps can be constructed. Two issues are worthy of consideration during design. First, for wye-connected windings, if possible, the tap winding should always be electrically placed adjacent neutral where the insulation level is lower, such that the insulation level of the on-load tap changer can be reduced to make the design more economical. For delta-connected windings or wye-connected winding the neutral insulation level of which is equal to line terminal insulation level, the tap winding is sometimes electrically sandwiched between two LV windings in such a way that the tap winding takes less transient voltages, so an on-load tap changer of lower insulation level can be applicable. Secondly, if possible, the tap winding should be located where leakage flux is small, i.e., outside of the main leakage flux channel between main windings; that way short-circuit forces and possible heating of the tap winding is reduced. Two types of tap winding are commonly used. One is a single winding with n sections; it can offer maximum (4n+1) tap voltages with bridging position, non-bridging position and reversing switch. It suits medium-size power transformers. Another is called coarse and fine tap windings, coarse winding has number of turns corresponding to n or $n + 1$ steps, and fine winding has n steps. It suits to large and very high voltage transformer.

By using the tap changing techniques mentioned previously, the voltage deviation can be reduced but not entirely eliminated. It is then necessary to increase excitation to have required voltage at HV terminals for a generation substation transformer to match the highest system voltage. This means that flux density in the core has to be increased. Such over-excitation needs to be checked on design stage, as shown in example that follows. The percentage voltage regulation approximates to Equation (1.16) when exciting current is neglected.

$$V_{regulation}\% = \alpha \left(R\% \cdot \cos\phi + X\% \cdot \sin\phi \right) + \frac{\alpha^2}{200} \left(X\% \cdot \cos\phi - R\% \cdot \sin\phi \right)^2 \quad (1.16)$$

Where α is number of times of rated current, $\cos\phi$ is load power factor, $R\%$ is percentage resistance voltage at rated load, $X\%$ is percentage reactance voltage; these are all per Equation (1.6)

$$R\% = \frac{LL}{rated\ KVA} \times 100; X\% = \frac{I_2 X_2}{V_2} \times 100 \quad (1.17)$$

Where LL is load loss, kW, X_2 is reactance seen from secondary side, Ohm, I_2 is secondary rated current, A, V_2 is rated secondary voltage, V. The deduction of the equation can be found in Reference [3].

Example 1.2

20/27/33 MVA unit, HV voltage is 145 kV, has tap range ±8 × 1.875%, the rated flux density in core is 1.66 Tesla. The load loss at 20 MVA is 46.547 kW. The percentage impedance at 20 MVA is 10.23%. The percentage resistance at 20 MVA is

$$R\% = \left(46.547 \times 100 \right) / \left(20 \times 1000 \right) = 0.233\%$$

The percentage reactance at 20 MVA can be calculated based on Equation (1.11)

$$X\% = \sqrt{10.23^2 - 0.233^2} = 10.227\%$$

At 33 MVA, the number of times of rated current is a = 33/20 = 1.65, for load with 0.8 lagging power factor, the percentage voltage regulation based on Equation (1.16) is

$$V_{regulation}\% = 1.65 \times \left(0.233 \times 0.8 + 10.227 \times 0.6 \right) + \frac{1.65^2}{200} \left(10.227 \times 0.8 - 0.223 \times 0.6 \right)^2$$

$$V_{regulation}\% = 11.31\%$$

HV system conditions are 145 kV nominal voltage with ±5% fluctuation, 60 Hz with ±5% range. In order to output top rating power of 33 MVA at 5% higher than nominal voltage and 5% less than rated frequency, the voltage of the transformer has to be increased by 11.31% + 5% + 5% = 21.31%. If there was no tap changer, the flux density would have to be increased to 1.66 × (1 + 0.2131) = 2.01 Tesla. This flux density drives the core into saturation, makes the unit unable to perform well. With tap changer and tapping up to maximum voltage position, the voltage is then increased by 8 × 1.875% = 15%. The rest, 21.31% −15% = 6.31%, is achieved by exciting the transformer further; the core flux density is then increased, 1.66 × 1.0631 = 1.76 Tesla. The core is not saturated at 1.76 Tesla, but the core temperature needs to be checked to ensure core insulation is not overheated.

1.1.6 WINDING DISPOSITION

Winding disposition affects quantities of materials used. The following example shows the difference in material cost of active parts between two real designs, while performances are similar to one another.

Example 1.3

Unit has rated power 60/80/100 MVA, LV is 20kV × 35 kV dual voltage wye connection with ±10% tapping range. HV is 115 kV delta connection. Two optional designs exist. Option 1 makes two LV windings, LV1 and LV2 as shown in Figure 1.6a and b. For 20 kV voltage, only LV2 winding is in the circuit; LV1 winding is not in the circuit so it doesn't take the load. For 35 kV voltage, both LV1 and LV2 windings are in circuit and connected in series. Option 2 is shown in Figure 1.6c and d. LV1 winding consists of two equal halves, for 20 kV voltage, the bottom and top halves of LV1 winding are connected in parallel, then are connected to LV2 in series. For 35 kV, the bottom and top halves of LV1 winding are connected in series, then connected to LV2 in series. In this design, unlike design option 1, LV1 winding shares the load at 20 kV connection, so LV2 winding can be smaller. If LV2 winding has smaller radial build, it makes HV winding smaller, too. As the result, the copper weight of windings in option 2 is less than one in optional 1, as shown in Table 1.1.

When transformers have two sets of LV terminals to supply power to two networks simultaneously, two design options may exist. Option 1 is that two LV windings are placed inside and outside HV winding respectively, which is usually called radial split. Option 2 is that the two LV windings are stacked one on the top of another, which is usually called axial split, as shown in Figure 1.7. In some cases, both radial split and axial split designs meet the same requirement such as insulation level, impedance, load loss and no-load loss, while axial split LV winding design has the advantage of material saving and size reduction. It should be noted that the axial split design fits only some cases and needs relevant manufacturing technology.

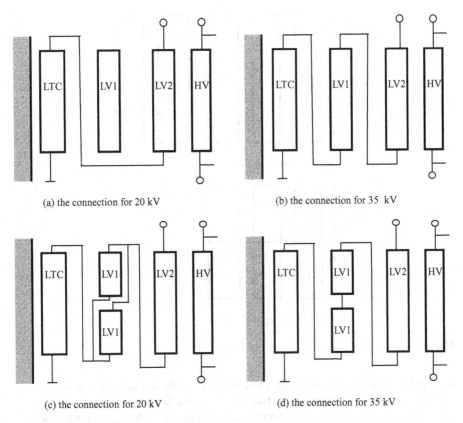

(a) the connection for 20 kV   (b) the connection for 35 kV

(c) the connection for 20 kV   (d) the connection for 35 kV

FIGURE 1.6   Two types of winding disposition.

TABLE 1.1
Active Part Material Cost and Performances Comparison

	Copper Weight	Core Steel Weight	Total Cost	Impedance (%)		No-Load Loss	Load Loss (kW)	
Design	(kg)	(kg)	($)	20 kV	35 kV	(kW)	20 kV	35 kV
Option 1	16127	36762	255,430	7.7	9.0	43.3	133	123
Option 2	14551	37434	243,267	8.06	8.65	44.3	124	125

Note: the costs per kilogram are assumed to be $9.0/kg for copper, $3.0/kg for core steel.

1.1.6.1 Winding Space Factor

The definition of winding space factor is the ratio of the copper area of winding to the core window area. It is one of economic measurement to see if a design is material economically. The more window area is occupied by winding conductors, the closer

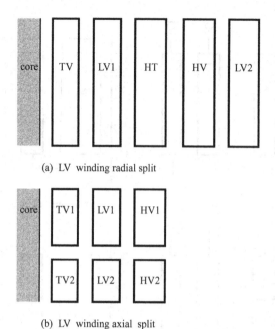

(a) LV winding radial split

(b) LV winding axial split

FIGURE 1.7   Radial split and axial split windings.

the winding space factor is to 1, the more economical the design is. The winding end insulation clearances to top and bottom yokes, the barrier which separates low voltage winding and high voltage winding in same phase and radial gaps between windings in different phases are mainly decided by transformer insulation levels, and need space in the window area. The radial spacers, between sections in disc winding or between turns in helical winding, also affect the space factor. The thicker the spacer is, the lower the space factor. On the other hand, a thinner spacer makes the space factor high, but oil flow may be slow down or blocked which may cause heating issue.

1.1.7 WINDING CONNECTIONS

Nearly all electric power is transmitted through three phases. Transmission can be achieved either by a three-phase transformer in which each phase of magnetic circuit is interlinked through one core, or by three single-phase transformers, the cores of which are independent. The most common connection types of the three-phase are wye and delta. The delta connection offers high current capability while the wye connection offers high voltage capability. Three connections of both windings are discussed in this section.

1.1.7.1 Delta–Delta Connection

When each phase, or each single phase transformer, is identical, such as identical impedance, and load on each phase is balanced, there is nothing to differentiate one phase from another except the phase displacement of 120° in both voltage and current between phases at power frequency. Assuming load currents are sinusoidal, the line currents and voltage have relation with phase current and voltage as follows:

$$I_{line} = \sqrt{3} \cdot I_\Phi; V_{line} = V_\Phi \qquad (1.18)$$

Where I_{line} is line current, I_ϕ is phase or winding current, V_{line} is line-to-line voltage, V_ϕ is phase or winding voltage. The current and voltage on the secondary side have the same relations. It should be noted that the balanced currents in each phase is achieved by equal impedance of each phase. In delta–delta connections, there are two parallel paths between each pair of terminals as shown in Figure 1.8; the line currents are divided among these parallel paths based on path impedance. If each phase has different impedance, the transformer or transformer bank cannot deliver its full power without overloading the phase having the smallest impedance.

1.1.7.2 Wye–Wye Connection

With balanced phase voltages, the line-to-line voltage and current have the relations with phase voltage and current as in the following equation:

$$V_{line} = \sqrt{3} \cdot V_\Phi; I_{line} = I_\Phi \qquad (1.19)$$

The definitions of parameters are the same as before. The neutral current of primary side and secondary side, I_N and I_n, are

$$I_N = I_{line,A} + I_{line,B} + I_{line,C}; I_n = I_{line,a} + I_{line,b} + I_{line,c} \qquad (1.20)$$

When there is a neutral line at primary side, each primary phase current is independent, it means that each phase current flows through primary winding and returns through the neutral line, not interfering with other phase current. In such a case, each phase can be loaded independently, therefore a single phase load can be connected

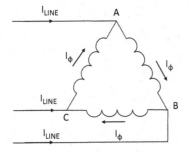 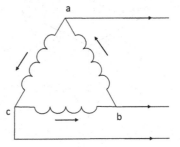

FIGURE 1.8   Delta–delta winding connection.

between any phase and secondary neutral. When transformer primary neutral is isolated, the sum of instantaneous current in each phase must be zero.

$$i_{line,A} + i_{line,B} + i_{line,C} = 0 \qquad (1.21)$$

When the line-to-line voltage applied to the primary terminal is balanced, and if the three phases have the same excitation characteristics, the exciting current and line-to-neutral voltage of each phase are balanced, as shown in Figure 1.9c. If the excitation characteristics of each phase are different—for example, phase B needs less exciting current than other two phases, then it receives more exciting current forced by other two phases, and consequently its phase-to-neutral voltage is more than its normal balanced voltage, as shown in Figure 1.9d. As a result of such unbalanced excitation, the neutral shifts from its geometrical center, meaning the neutral voltage to ground is not zero. Another factor causing neutral shift is single phase-to-neutral load on the secondary side, details of which are discussed in Chapter 10.

1.1.7.3 Delta–Wye and Wye–Delta Connections

Step-up or GSU transformers are usually delta primary and wye secondary. Step-down transformers can be either wye–delta, or delta–wye, depending on which side needs neutral. The delta connection balances the wye voltages to neutral, while the

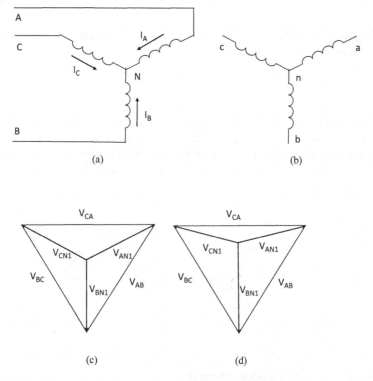

FIGURE 1.9   Balanced and unbalanced exciting of each phase.

wye connection balances the currents in delta. This balancing action sometimes is an advantage of delta–wye and wye–delta connections, since three phases take nearly equal shares of a balanced three-phase load, even when their impedance and excitation performance are unequal [1]. Another popular connection is auto connection, it will be discussed in detail in Chapter 10.

1.2 PRACTICAL CONSIDERATIONS IN DESIGN

The constant challenge in design is to find the most suitable design, both technically and economically. Due to complexity of design and requested guarantees of performances, a satisfactory result cannot be obtained simply by solving a series of equations. The design process is basically to find the most suitable design by repeating calculations with different parameters. Several parameters are adjustable during the optimizing process; here we list some of the more important ones.

The current density in conductor affects directly I^2R loss as well as cooling of winding. The maximum current density in winding cannot be too high due to cooling method limitation. The values of the current density, for Class A insulated transformers, vary from about 3.2 A/mm^2 for distribution transformers to 5.5 A/mm^2 for larger transformers with forced cooling [4]. The maximum current density of 4.0 A/mm^2 for larger transformers with non-forced cooling is also suggested [5]. For compact design such as mobile power transformer, 7~7.3 A/mm^2 are selected for 75~95°C rises, and Class H insulation such as NOMEX has to be used as wire insulation.

The flux density in core is another key design parameter. The maximum flux density is limited by core saturation, inrush current and noise generation. When unit is not overexcited in service, the flux density may be selected around 1.7 Tesla, when network could cause over excitation of unit, such over excitation condition should be known before normal or rated flux density is selected. When core sound level is requested to be very low, low flux density is used in design. However, reducing the flux density below certain level doesn't help further reduction of core sound level, because low flux density usually brings bigger core the contribution of the core weight to the sound level increases. Detail discussion is conducted in Chapter 9.

The relation between current density and flux density, for the case of minimum I^2R loss ($J_1 = J_2$), is [4]

$$J = 10.59 \times f \frac{B_m}{k_i} \cdot \frac{A_{Fe}}{T_{mean}} \cdot \left(\%LL\right) \tag{1.22}$$

Where J is current density, A/mm^2, f is frequency, Hz, B_m is maximum flux density, Tesla, Supposing
$k_i = 1 + Eddy_1 \% \approx 1 + Eddy_2 \% = 1 + Eddy_{average}\%$, is average winding loss factor, here it is assumed that the eddy current losses of both windings are same. A_{Fe} is net cross sectional area of core, m^2. T_{mean} is average of mean turn of winding 1 and winding 2, m. $\%LL = LL_\phi / S_\phi$, where LL_ϕ in kW and S_ϕ in MVA are load loss and rated power per phase respectively. Copper resistivity $\rho = 2.096 \times 10^{-9}$ Ω·m at 75°C is used. If J and B_m are chosen independently, the transformer will have a natural value of load loss depending on the ratio of A_{Fe}/T_{mean}. If the losses are guaranteed, the choice of J must correspond to that of B_m and possible values of A_{Fe}/T_{mean}.

1.2.1 Minimum I^2R Loss

Minimum I^2R losses can be obtained when the current densities in each winding are same [4]

$$J_1 = J_2 \tag{1.23}$$

Where J_1 and J_2 are the current densities in winding 1 and winding 2 respectively. Winding eddy loss can be kept quite as small as 5% of I^2R loss in small transformers, 15% in large units in good designs.

1.2.2 The Most Economic Utilization of Active Materials

Based on reference [4] it is found that the most economic utilization of core and winding materials occurs within a limited range of winding shape ratio. Winding shape ratio is defined as winding height / average of mean turn (i.e., average circumference of winding) of LV and HV windings, H_{wdg}/T_{mean}. When the shape ratio is in a range from 0.3 to 1.0, it is very approximately correct for small transformers. As the size of transformer increases, the range of H_{wdg}/T_{mean} for practical units narrows down. The majority of transformers with normal values of losses and reactance, H_{wdg}/T_{mean} lie between 0.55 and 0.75, as shown in Figure 1.10a, Area 1. When the designs are only based on minimum cost of active material which is discussed below, H_{wdg}/T_{mean} lie between 0.6 and 0.7, it is Area 2 in Figure 1.10a. For large transformers, height restrictions due to transport difficulties limit the value of H_{wdg}/T_{mean} to less than about 0.6 in the majority of units. Investigation results of practical designs are shown in Figure 1.10b, they include various designs such as 2-winding to multiple-winding, radial split and axial split, auto transformers etc., each transformer has its own guaranteed performances like no-load loss, load loss, impedance, sound level even overload temperature limits, these may be the reasons why some of designs are out of the limit range mentioned in reference [4], but both of trends are the same. Figure 1.10b also indicates that besides guaranteed values, same attentions are needed on material utilization. Effective use of materials to achieve the same guaranteed performance is the goal of a good design.

1.3 ACTIVE PART MATERIAL COST

1.3.1 Loss and Mass Ratio for Maximum Efficiency

The maximum efficiency happens at a fractional load, k_l, in which load loss is equal to no-load loss

$$k_l^2 \cdot LL = NLL \tag{1.24}$$

Where load loss is $LL = m_{Cu} \times ll$, no-load loss is $NLL = m_{Fe} \times nll$, m_{Cu} and m_{Fe} are mass of copper and core steel respectively, ll and nll are specific load loss per unit weight

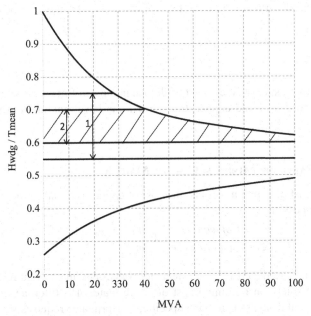

(a) Practical limits of economic design of active part by Reference [1]
    Area 1: most design with normal losses and reactance
    Area 2: design based only on minimum cost of active parts

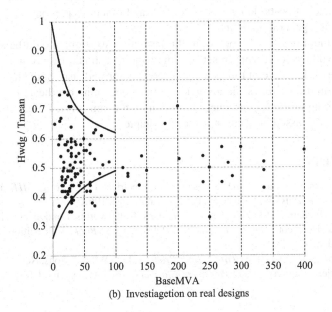

(b) Investiagetion on real designs

FIGURE 1.10   Relation of H_{wdg}/T_{mean} with **MVA** for economical design.

and no-load loss per unit weight respectively. Equation (1.24) thus can be re-written as

$$\frac{m_{Fe}}{m_{Cu}} = \frac{ll}{nll} \cdot k_l^2$$

(1.25)

If average working load or percentage of full load in most of time, k_l, is known, using above ratio of core steel and copper, transformer can have maximum efficiency in most of time.

1.3.2 MASS RATIO FOR MINIMUM COST OF MATERIAL

For minimum cost of active materials, the cost of copper is equal to the cost of the steel [4], or the cost ratio

$$\frac{cost\ of\ copper}{cost\ of\ core\ steel} \approx 1;\ \frac{m_{Fe}}{m_{Cu}} \approx \frac{c_{Cu}}{c_{Fe}}$$

(1.26)

Where, c_{Fe} and c_{Cu} are the cost of core steel and copper per unit weight respectively. A design based solely on minimum cost of active material is often an impractical design, since a real design has to meet specified performance requests such as load loss, no-load loss, impedance, sound level, etc. These specified requests could make each design hard to meet the relation expressed in Equation (1.26). It should also be noted that the total cost of a transformer consists of not only the active part cost, but also other material costs, labor cost, overhead and others. Equation (1.26) can be treated only as a general guideline.

To make the transformer work well, the insulation structure should be suited to any type of voltages occurring in service, the windings should be strong enough to take any type of short-circuit forces without deformation and also should have a good cooling system to prevent insulation material from becoming overheated and accelerating the aging process, consequently shortening the unit service life, All of those topics will be discussed in detail in following chapters.

REFERENCES

1. Members of the staff of the Department of Electrical Engineering, *MIT, Magnetic Circuits and Transformers*, The MIT Press, Cambridge, 1962.
2. Bernard Hochart, *Power Transformer Handbook*, Butterworths, London, etc., 1987.
3. Martin J. Heathcote *Heathcote, The J & P Transformer Book*, 13th edition, Newnes, Amsterdam, etc., 2007.
4. R. Feinberg, *Modern Power Transformer Practice*, John Wiley & Sons, New York, 1979.
5. Jim Fyvie, *Design Aspects of Power Transformers*, Arima Publishing, UK, 2009.

2 Core

Based on Faraday's law, the voltage induced in a winding has a relation with flux going through the area it circles and its number of turn as

$$e = -N\frac{d\phi}{dt}; \phi = \int_s B \cdot ds \qquad (2.1)$$

Where e is induced voltage (electromotive force) in Volt, N is the number of turns in the winding, ϕ is the flux in Weber; t is time in Second; B is flux density in Tesla; S is the area the flux is going through in m^2. As can be seen, higher induced voltage requires more flux. The engineering solution to increase flux is to use iron core, in such a way the flux density can be increased so much higher than in air, that even a smaller area can hold much more flux. As a result, the higher induced voltage is achieved, while windings sizes as well as the overall dimension of transformer are reduced. The second benefit of iron core is lower load loss due to smaller windings. The third benefit is, unlike the air core, the flux is almost totally confined in the iron core, which reduces stray loss caused by stray flux and less coupling with the adjacent circuit. The undesirable effect of iron core is its loss, it reduces efficiency and contributes to temperature rise of oil and the core itself. The nonlinear magnetizing characteristics of iron core cause the waveform of exciting current to be non-sinusoidal when the flux varies sinusoidally which will be discussed in this chapter. First, it is necessary to know about core materials.

2.1 CORE MATERIALS

Two types of core steel are commonly used: hot-rolled non-oriented, and cold-rolled grain-oriented. The processes of hot-rolled non-oriented steel produces lamination sheets in which grain (crystals) were almost randomly oriented [1]. As the consequence, its magnetization properties and losses are essentially identical in any direction of magnetic field applied. This process requires the silicon content to be limited to 4.5%, in order to avoid an unworkable degree of brittleness. The resistivity of the lamination is about 60×10^{-8} $\Omega \cdot m$.

The cold-rolled process orients the grains in one direction. The overall effect is that first, core operating flux densities can be increased and still give a substantial reduction in core loss over the non-oriented hot-rolled steel. Secondly, the cold rolling gives a smoother surface finish and thereby improves the space factor [1]. Present electrical core steels have a silicon content of about 3%, which gives a resistivity about 47.2×10^{-8} $\Omega \cdot m$. Higher silicon content could offer a better magnetic behavior; for example, with 6% content, the magnetostriction of the steel disappears. Magnetostriction, a change in length of the lamination caused by induction, contributes to noise of the transformer core. Also, higher silicon content can produce higher resistivity, which could further reduce eddy current loss in the steel. However, the brittleness also increases with silicon content, making it difficult to roll them in the mill as well as cut and stack in later core making process [2]. The magnetic behaviors and losses of this steel are strongly dependent on the direction of magnetizing. The material has its best magnetic behaviors along the rolling direction, and a well-constructed core should take this advantage.

It should be noted that, in order to reduce eddy current loss, the grain or crystal sizes have to be reduced. Scribing lines on the lamination by laser or by mechanical method could refine the grain domain size by providing local stress. However, after scribing, it is not suggested to anneal the core, because this could release the local stresses. The loss of cold-rolled grain-oriented steel in term of watt per kilogram is nearly half of hot-rolled non-oriented steel. This means that a transformer could be built smaller as well as have lower loss with cold-rolled steel than with hot-rolled steel.

A relatively new core steel is amorphous steel [3]. The amorphous metal is not crystalline like silicon steel. Due to non-crystalline molecular structure, the amorphous metal core exhibits very low hysteresis losses. The randomness of molecular structure of amorphous metal causes less friction than silicon steel when a magnetic field is applied. The resulting reduction in hysteresis losses makes amorphous metal core transformers have no-load loss 3~7 times lower than silicon steel core transformers. Because of very thin, usually 1 mil thickness compared to 7~12 mil of silicon steel, an amorphous metal core has lower eddy current loss, too. However, there are issues which have to be dealt with in application, the first of which is extreme hardness and thinness. Amorphous metal is four to five times as hard as silicon steel. This characteristic causes cutting tools with carbide tips to wear out 1000 times faster than when used on silicon steel. The design has to involve minimizing the cutting of the metal. The metal is very brittle due to its thinness, and any type of bend should be avoided in production. The second issue is a lower space factor caused by the thinness. It has about 80% space factor compared to 97% for silicon steel. Because of the lower space factor and lower saturation induction level of 1.5 Tesla [2], amorphous metal cores require more material than conventional cores. Third, ferroresonance in the transformer occurs when the inductive reactance of the transformer resonates with capacitive reactance of the feeder. The resonant condition can result in overvoltage. A high no-load loss transformer could attenuate the ferroresonant overvoltage. With very low no-load loss like amorphous unit, ferroresonant overvoltage is not attenuated. The solution to this problem is to apply an arrester, which incurs additional cost. No large transformer has yet been built using amorphous metal due to difficulties of assembling large lamination.

Good surface insulation coating keeps eddy current loss to a minimum. It should be noted that the insulating coating can prevent currents from flowing between laminations, but it does not prevent the induced voltages from being developed in the laminations. The induced voltage is proportional to the plate width [4]. With increase of the core cross-sectional area, the lamination plate width increases, and the higher voltage is induced. When there are burrs at lamination edges caused by dull cutting tools, and number and length of the burrs of one lamination are enough to touch the burrs of the next lamination to make a loop, a current driven by the induced voltage circulates in the loop and generates losses, which will be discussed later. One way to reduce this voltage is subdividing the cross-sectional area so that the plate width is reduced as well as the induced voltage. Another way is applying additional insulation coating. High-quality core steels have good insulation coating, and by using them, it is not normally necessary to provide additional coating regardless of the size of the core.

Core is supported and clamped tight by a core clamping structure. The insulation between the core and the clamping structure is made by pressboard when the core surface temperatures are lower than 105°C, or by Nomex or its equivalent when the core surface temperatures are higher. The lock plates which hold top and bottom clamp frames in place are usually carbon steel when the leakage flux hitting on it is low. When this radial leakage flux density is high, the eddy current loss in the lock plates can raise the temperature significantly. Dividing the lock plate into several strips or application of stainless steel can reduce temperature rises; as a tradeoff, the mechanical strength of the lock plate is weakened.

2.2 CORE TYPES

For the three-phase core form transformer, two types commonly used are the three-leg core and the five-leg core.

2.2.1 THREE-LEG CORE

The three-leg core form has a unique feature: there is no return path for the magnetic flux converging in the top and the bottom yokes from the three legs. For positive or negative sequence fluxes generated by balanced applied voltages, the flux through each leg has the same magnitude and is 120° apart from each other; therefore, the vector sums of these fluxes in the yokes are zero [5]. For residual zero sequence flux produced by unbalanced applied voltages, it has to leave the core from one yoke and returns to another yoke through oil. The path in the oil has high reluctance due to the oil permeability being much lower than the core steel's. This high reluctance path impresses the flux and reduces the voltage imbalance. For positive and negative sequence flux, the three-leg core is a closed circuit; for zero sequence flux, it is an open circuit.

2.2.2 FIVE-LEG CORE

For large three-phase units, the three-leg core may cause shipping height over the limits of tunnels and bridges. In such cases, the five-leg core is used to reduced yoke height, therefore the core height and the overall height of transformer are reduced; to offset this, the overall length is increased by two unwound legs. Another advantage of the five-leg core is that the cooling of yokes is greatly improved without providing special ducts. The changes of cross-section at the joint between legs and yokes cause flux to transfer from step to step across the plane of lamination, thus increasing the eddy current loss. The flux densities in various paths of a five-leg core depend on the relative reluctances of the paths in the core, so it is difficult to determine their proper relative cross-sectional areas. Experience has shown that if the main portion of the yoke is made about 58% of the wound leg cross-section, and if the unwound leg has 40~50% of it, then the flux densities in various paths will be substantially equal, and a reasonably low core loss will be achieved [1].

Another occasion on which the five-leg core is needed is when zero-sequence impedance is requested, having similar magnitude as positive sequence impedance. For both positive/negative sequence fluxes and zero sequence flux, the five-leg core is a closed circuit, because the unwounded legs offer a low reluctance path to zero sequence flux as the wound legs to positive/negative sequence fluxes.

2.3 NO-LOAD LOSS

When the transformer is energized, the core steel and the circuit through which exciting current circulates produce loss called no-load loss. The watts consumed by the core are in the form of heat, which contributes to core and oil temperature rises. The no-load loss is always produced when the transformer is energized regardless whether it carries a load or not. No-load loss is usually guaranteed below a limit. The no-load loss consists of several components discussed in this section.

2.3.1 COMPONENTS OF NO-LOAD LOSS

The loss is produced due to core steel material and core making production.

2.3.1.1 Hysteresis Loss

This loss is from core material and is proportional to the area of the hysteresis loop. When a metal is subjected to an alternative magnetic field, its atoms reorient themselves to be in alignment with the magnetic field and form tiny magnetics. The movement causes molecular friction that is given off in the form of heat and is called hysteresis loss [3]. Hysteresis losses occur at very low, essentially DC, frequency, thus in loss separation studies, the hysteresis losses can be measured independently by testing at low frequency [2].

2.3.1.2 Eddy Current Loss

It is also from core material, generated by eddy currents which are produced by induced voltage in lamination in response to an alternating flux. There are two types of eddy current loss in core steel. One is classical eddy current loss, which relates to thickness and resistivity of lamination steel. Another relates to the movement of domain wall, which is explained in detail in reference [2] and is called non-classical eddy current loss. This loss is significantly higher than the classical eddy current loss. The non-classical loss depends on the size of the domain: the larger the size is, the greater the loss is. In order to reduce such non-classical eddy current loss, the domain size has to be reduced, which can be achieved in practice by laser or mechanical scribing of the lamination. After scribing, the loss is reduced by about 12%.

2.3.1.3 Additional Losses [6,7,8]

For grain-oriented steel such as cold-rolled steel, the minimum losses occur when the flux direction is the same as the rolling direction. When these two directions differ, the loss and the magnetizing power increase, with the magnetizing power increasing more significantly. In the following cases in practical cores, the flux has to change the directions and additional losses are generated. These losses also strongly depend on the type of material; they are highest in domain refined steel, lowest in regular core steel.

In the region near the holes in legs and yokes, the holes are needed for alignment during core stacking, and the fluxes have to go around the holes. The magnetic field in the core tries to maintain a minimum energy, so when the fluxes change their directions around the holes, they keep this path even after they passed the holes, because it costs additional energy to change back. As the result, the flux density around the hole area is a bit higher than the flux density far away from holes, this local high-flux-density area could generate a bit more loss.

In the corners of core, besides the fact that fluxes have to divert their directions away from the lamination rolling direction in the corners, causing extra loss and magnetizing power, the fluxes have to go through the oil gap to reach to another lamination at the joint. The oil gaps always exist in real core joints. Going through the oil gap also costs extra magnetizing power to drive the flux through low-permeability oil and extra loss. These losses are called joint loss.

Different joint patterns make different gap sizes; the larger the gap size is, the greater the loss is. Measurements show that compared to changing lamination sheet joint patterns from layer to layer, the loss is 5~10% higher when the patterns are changed after every other layer; 10~15% higher for changing the pattern after every three layers, and 15~20% higher when the pattern is changed after every four layers.

2.3.1.4 Interlaminar Loss

There are two types of interlaminar loss. One is caused by leakage current penetrating through the lamination stack due to imperfect insulation coating of lamination. It is proportional to the square of the lamination width and inversely proportional to the mean value of insulation resistance per unit area through the core [1,2]. Another is caused by burrs on the lamination edges which touch adjacent laminations, making electrical loops for current to flow. As a result, extra loss is generated. Both kinds of losses are discussed in the following section.

2.3.2 Calculation of No-Load Loss

Hysteresis loss is proportional to applied frequency and maximum flux density of 2~2.5 power. Eddy current loss is proportional to square of applied frequency and square of maximum flux density. There are also additional loss, caused by holes and corner joints, and interlaminar loss. Besides all of these losses, there is also the exciting current which goes through primary winding produces I^2R and eddy current losses. Voltage applied on the insulation structure also produces so-called dielectric loss. However, both exciting current and leakage current in the insulation structure are very small, and their losses are negligible. The total no-load loss, NLL, is

$$NLL = P_{hysteresis} + P_{eddy} + P_{additional} + P_{interlaminar} \qquad (2.2)$$

Where

$$P_{hysteresis} = k \cdot f \cdot B_{max}^n; \; P_{eddy} = (1/24) \cdot \gamma_{core} \cdot \omega^2 \cdot d^2 \cdot B_{max}^2 \cdot Vol_{core} \qquad (2.3)$$

Where $P_{hysteresis}$ is hysteresis loss, P_{eddy} is eddy current loss, $P_{additional}$ and $P_{interlaminar}$ are additional loss and interlaminar loss respectively. k is a constant depending on material, n is the Steinmetz constant having a value of 1.6 to 2.0 for hot-rolled lamination and a value of more than 2~2.5 for cold-rolled laminations due to use of higher operating flux density [9], f is frequency, γ_{core} is core steel conductivity, ω is angular frequency, d is the thickness of individual lamination, B_{max} is the peak value of flux density, Vol_{core} is the volume of core. In the practical designing process, the no-load loss is not usually calculated in this way; instead, the specific loss (W/kg) of core steel is used and influential factors are taken into account. The no-load loss is calculated as follows.

$$NLL = (W / kg) \times W_{core} \times F_{build} \times F_{destruction} \times F_{freq} \times F_{temp} \qquad (2.4)$$

Where NLL is the no-load loss at 20°C; (W/kg) is specific loss which is different with different flux density and different type of steel, W_{core} is core weight in kg, F_{build} is building factor which is about 1.15 for a well-designed grain-oriented steel core, it counts the additional loss from joints and corners, $F_{destruction}$ is the destruction factor to count additional loss caused by holes in lamination, F_{freq} is the frequency factor, which is $F_{freq} = (f/60)^{1.5}$, where f is frequency, Hz; F_{temp} is temperature factor to count the loss variation with the temperature, which is $F_{temp} = 1 + [(20 - T) \times 0.00065]$, where T is a temperature other than 20°C.

Example 2.1

A unit's core weight is 74,041 kg, the rated flux density is 1.68 Tesla, the specific loss under this flux density and 60 Hz is 1.413 W/kg, $F_{build} \times F_{destruction} = 1.18$, The no-load loss at 20°C, 60 Hz is calculated as NLL = 1.413 × 74,041 × 1.18 × 1.0 × 1.0 = 123.45 kW.

As can be seen from Example 2.1, the flux density directly affects no-load loss. Rated designed flux density is generally in the range of 1.6~1.7 Tesla. To achieve low no-load loss, low flux densities such as 1.35 to 1.6 Tesla are usually selected. These low no-load loss units often have high load loss because more winding turns is required to reduce volt per turn and flux density. The recent trend is that higher flux densities are selected in designs to save core steel and winding copper; in such cases, the flux density increments needed to overcome voltage regulation, system voltage and frequency fluctuations in real operation situations have to be carefully studied to avoid core saturation or overheated.

2.3.2.1 Interlaminar Losses

The interlaminar loss produced by bad surface coating insulation, P_{coat}, is estimated using another format of Equation (2.3) as.

$$P_{coat} = \frac{1}{24} \cdot \gamma_{coat} \cdot \omega^2 \cdot d^3 \cdot B_{max}^2 \cdot w \qquad (2.5)$$

Where w is lamination width, γ_{coat} is the conductivity of the surface coating, B_{max} is the peak value of flux density, ω is angular frequency. The loss, P_{coat}, rises with low resistance of the surface coating, and higher-than-expected core loss might mean lower insulation resistance of the coating. With adequate insulation, this loss is about 1~2% of total no-load loss. In the case that core lamination is too wide and it is worried that the surface coating does not have sufficient insulation, the widest lamination sheet can be split into two and separated by a cooling duct in between. Another option is additional insulation coating, but this makes core stacking factor poor, and it means that the effective core area is further reduced by the additional insulation coating.

When cutting tools are dull, burrs are made at the slit and punched edges. When the burrs are long enough to reach adjacent lamination edges, electric loops between the laminations are formed which offer circuits for currents to flow. For estimation purposes, considering the worst situation in which the surface of every step of a core is short-circuited by the burrs, the sum of voltages exerted at each step is volts per turn (V/N), the sum of currents flowing in each step is $I = (V/N)/R_{core\ surface}$, The loss generated, P_{burr}, is then.

$$P_{burr} = F_{burr} \cdot \frac{(V/N)^2}{R_{core\ surface}} \qquad (2.6)$$

Where $R_{core\ surface}$ is total surface resistance of the core, F_{burr} is the factor considering approximation in the estimation. In a practical situation in which the laminations are cut very badly, this loss went up as high as about 30% of total no-load loss.

With high-quality core steel and under quality controlled production line, these two types of losses are generally much smaller compared to the no-load loss in Equation (2.4). The purpose of discussing these interlaminar losses is to offer a clue to investigation on high no-load loss tested, and quality control of the core production. Burrs on lamination edges can cause no-load loss much higher than expected in worst cases. The measure to reduce this loss is to check the cutting tools on a regular basis to ensure they have the required condition. With HiB steel and carbide-steel tools, burr heights of less than 0.02 mm are possible.

2.4 EXCITING CHARACTERISTICS[4,5,9,10]

2.4.1 CORE EXCITING CURRENT

Two basic physical laws are used here. First, Faraday's law of induction, the induced voltage (or electromotive force), $v(t)$, by magnetic field is described in Equation (1.1). For convenience, it is repeated here:

$$v(t) = -N(d\phi/dt) \qquad (2.7)$$

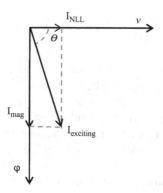

FIGURE 2.1   Vector diagram of exciting current.

Where N is number of turns in winding, ϕ is the magnetic flux, t is time. Based on the law, first the induced voltage is only related to the flux and the number of turns. Secondly, if the flux is sinusoidal, the induced voltage is also sinusoidal but 90° leading the flux. If applied voltage is sinusoidal, the flux produced is of sine wave, with 90° lagging while exciting current is non-sinusoidal.

The flux, ϕ, is produced by the magnetomotive force of the exciting current. The exciting current has two components as shown in Figure 2.1, one is for core loss, I_{NLL}, another is for magnetizing core, I_{mag}, called magnetizing current. The exciting current, $I_{exciting}$, is then.

$$I_{exciting} = \sqrt{I_{mag}^2 + I_{NLL}^2} \tag{2.8}$$

For core steel used at present, the relation of the flux with exciting current is like Figure 1.2, which is nonlinear; the core steel has variable permeability instead of a constant one like air. Because of such magnetic nonlinearity of the core material, when the flux is of sine wave, the exciting current is non-sinusoidal. A measurement of applied voltage and exciting current is shown in Figure 2.2, and a graphical solution based on sinusoidal flux wave and relation between flux and exciting current of a core material is shown in Figure 2.3. The exciting current is peaky wave containing third harmonic components. On the other hand, if the exciting current is sinusoidal in nature, the flux wave will be flat-topped, and the induced voltage in windings will be peaky containing third harmonic components, the content of which depends on type of magnetic circuit (separate or interlinked) and type of winding connections (wye, delta or zigzag); these will be discussed later.

It should be noted that even through the exciting current is non-sinusoidal, the calculations related to it are conducted assuming it is sine wave. These calculations generally give good engineering accuracies because, first, the resistance in primary circuit is small, and second, the exciting current is very small for modern core steels, i.e. usually less than 1% of rated current. Both make the voltage drop that they caused negligible.

The second law used here is Kirchhoff's current law: the instantaneous sum of the currents flowing to and from a common junction or star point is zero.

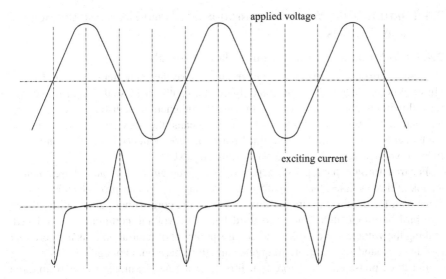

FIGURE 2.2 Typical wave shapes of applied voltage and exciting current.

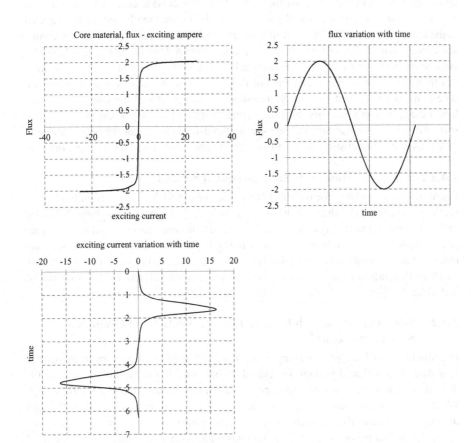

FIGURE 2.3 Graphical solution of exciting current.

2.4.2 INFLUENCE OF WINDING CONNECTIONS ON THIRD HARMONIC VOLTAGES AND CURRENTS

2.4.2.1 Y–Y Connection with Both Isolated Neutrals

The third harmonic components in each phase have the same magnitude and are in phase. When the neutrals are isolated, there is no path for these third harmonic currents flowing in phases and in lines, making the magnetizing current almost sinusoidal and the flux wave flat-topped, which means the flux has third harmonic components. The amounts of the third harmonic components in the flux and in its induced voltage depend on situations discussed next.

For transformers having an independent magnetic circuit for each phase, such as a bank of three single-phase transformers, the magnetic circuit provides a low reluctance path for the third harmonic flux to flow in the core without attenuation. When the third harmonic flux flows, the third harmonic voltage component is induced, making the voltage wave peaky; the voltage potential of neutral point is not zero, but oscillates around zero at triple frequency and third harmonic voltage.

For three-phase three-leg core transformers, the third harmonic fluxes from three legs cannot cancel each other out in the yokes because they are in phase; they have to leave one yoke and return to another yoke from the outside core, which is a high reluctance path. As a result, the third harmonic fluxes are greatly reduced compared to the bank of three single-phase transformers, so as the induced third harmonic voltage component (2~5% of fundamental voltage compared to 30~70% of fundamental voltage in single-phase units). The directed advantage is greater stability of the neutral than single-phase units. Because of this, it is possible for three-phase three-leg core units to take a moderate single-phase load, saying 10% of the total three-phase load rating, between one line and the neutral without causing an excessive neutral shift. The main disadvantage of three-phase three-leg core is that the third harmonic flux causes additional stray losses in structural parts on its path outside the core.

2.4.2.2 Y–Y Connection with Both Grounded Neutrals

There is a path available for the third harmonic current to flow; the magnetizing current is peaky, and the induced voltage is sinusoidal. Its main disadvantage is that the third harmonic current may cause interference in communication circuit running parallel to power lines. If a tertiary delta winding is provided, the third harmonic currents will be reduced but not completely eliminated. The current shared by the tertiary winding depends on the relative values of impedance offered by the two paths. Reference [10] gives more detailed analysis.

2.4.2.3 Y–Y Connection with Isolated Primary Neutral and Grounded Secondary Neutral

In a shell type or bank of three single-phase transformers, when the primary neutral is isolated, the induced voltage in secondary has third harmonic components. This third harmonic voltage can be measured between the secondary neutral and ground. When the secondary neutral is grounded, a third harmonic current can flow through the circuit. It seems that in such a situation, the third harmonic voltage can be suppressed. But that is not totally true; it depends on the phase relation between the third harmonic current in secondary winding and sinusoidal current in primary winding. Reference [4] gives a detailed study on the topic.

2.4.2.4 Y–D or D–Y Connection

The third harmonic voltages in each phase are consumed by driving third harmonic currents circulating around the delta. As the result, there is no third harmonic voltage, and no third harmonic current in the line current.

2.4.3 UNDESIRABLE FEATURES OF THIRD HARMONICS

The existence of third harmonic current has very undesirable features. First, the circulation of third harmonic current in winding may overheat the winding. This issue may become serious when the transformer primary windings are connected in an interconnected star, and the generator and the transformer neutrals are jointed together. Secondly, as mentioned earlier, third harmonic current flowing in power lines may interfere the communication line nearby. Third, in the case of a bank of three single-phase transformers of wye-wye connection, it has been proved experimentally that a fourth wire connection on the primary side between the transformer bank and generator neutrals (which allows the circulation of the third harmonic currents) results in increasing no-load loss of the transformer to 120% of the obtained with neutral disconnected. In practice, the no-load loss has been found to reach three times the normal no-load loss of the transformer, the apparatus failed as a consequence [4].

Similar to the third harmonic current, the existence of third harmonic voltages puts additional stresses on winding to ground insulation structure. Distribution transformers have an usually large safety margin; it may not cause a problem, but it has considerable influence on the reliability of transformers having higher voltages. The third harmonic voltage may also cause electrostatic charging of adjacent lines and telephone cables, and possible resonance at third harmonic frequency of transformer windings and line capacitance.

2.4.4 CALCULATION OF EXCITING CURRENT

Voltage transformation is accomplished at the expense of an exciting current. There are two components in exciting current, the first of which is the power component which is consumed in no-load loss. This current, $I_{NLL}(\%)$ in percentage, can be calculated as follows.

$$I_{NLL}\left(\%\right) = NLL \, / \, S_n \times 100 \qquad (2.9)$$

Where NLL is no-load loss in Watts, S_n is rating of transformer in VA. Another component is magnetizing current which is reactive and produces the magnetization of core. In a practical designing process, the curve of VA/kg versus flux density of a given lamination material is used to calculate the magnetizing current.

$$I_{mag}\left(\%\right) = \frac{S_{mag}}{S_n} \times 100; \; I_{mag}\left(A\right) = \frac{\left(\frac{VA}{kg}\right) \times W_{core} \times F_{joint}}{\sqrt{3} \times V_{L-L}} \qquad (2.10)$$

Where S_{mag} is magnetizing power, W_{core} is core weight, F_{joint} is joint factor to count additional magnetizing current required at the joints, V_{L-L} is rated line to line voltage. The total current, which is called exciting current, is the vector sum of these two components as shown in Equation (2.8). Some features of exciting current are listed here:

- As mentioned earlier, the magnetizing current has harmonic components in order to induce sinusoidal voltage. These harmonic components are wattless.
- Although no-load loss and exciting current are lower in better-grade steel, such as HiB material (M0H, M1H, M2H), laser or mechanical scribed, compared to CGO material (M2, M3, M4, M5, M6, etc.), the saturation flux density has remained same (2.0~2.05 Tesla).
- In three-phase three-leg core form units, the exciting currents in outer legs are almost equal to one another, and their magnitudes are greater than the current in the center leg because reluctances of the outer legs are higher than the center leg's. To get the same flux density as in the center leg, the outer legs need more magnetomotive force, i.e., higher exciting current.
- Besides increasing no-load loss, joints need more exciting current to drive the flux through the oil gaps and change its direction. When the joints are loose with big gaps, the exciting current will be higher than tight joints with small gaps. Loose joints sometime result in higher sound level, too. So tested higher sound levels are sometime accompanied by tested higher exciting current.

Example 2.2

(2.1) A 150 MVA unit, HV line to line voltage is 161 kV wye connection. The calculated no-load loss is 69,900 Watts. The core weight is 54,029 kg. Under maximum flux density of 1.678 Tesla, the specific magnetizing power is VA/kg = 1.6. The magnetizing current is $I_{mag}(A) = (1.6 \times 54029 \times 1.0)/(\sqrt{3} \times 161000) = 0.31$ A.

The HV current at 150 MVA is $(150 \times 10^3)/(\sqrt{3} \times 161) = 537.9$ A.

The magnetizing current in percentage is $I_{mag}(\%) = (0.31/537.9) \times 100 = 0.058$ %.

The current due to no-load loss is $I_{NLL}(\%) = (69900/150 \times 10^6) \times 100 = 0.047\%$

The exciting current is $I_{exciting}\left(\%\right) = \sqrt{0.058^2 + 0.047^2} = 0.075\%$

(2.2) A 50 MVA unit, wye-connected LV L-L voltage of 21.6 kV is regulated through a regulating winding plus a PA, also the regulating winding is parallel connected to a series transformer in order to reduce the regulating winding current, as shown in Figure 2.4. At tap 15 L (19.56 kV L-L), all of main core, PA core and series core are magnetized. The exciting current under such situation is calculated next.

Main Core

The weight is 33,217 kg, volt per turn = 138.56. Under the maximum flux density of 1.61 Tesla, the specific magnetizing power is VA/kg = 1.336. The magnetizing current is $I_{mag,\ main}$ (A) = (1.336 × 33217 × 1.0)/($\sqrt{3}$ × 19560) = 1.3 A. The magnetizing power per phase is $S_{mag/\varnothing,\ main}$ = (1.3 × 19560)/($\sqrt{3}$ × 1000) = 14.7 KVA. The calculated no-load loss is 39,149 W.

Series Transformer Core

The weight is 3637 kg. Under the maximum flux density of 1.31 Tesla, the specific magnetizing power is VA/kg = 1.074. The magnetizing current is $I_{mag,\ series}$ (A) = (1.074 × 3637 × 1.0)/($\sqrt{3}$ × 1169.1 × $\sqrt{3}$) = 1.1 A The magnetizing power per phase is $S_{mag/\varnothing,\ series}$ = 1.1 × 1169.1/1000 = 1.3 KVA. Here, 1169.1 is the common winding voltage of the series transformer. The calculated no-load loss is 4050 W.

PA Core

There are two turns or one turn between taps in the regulating winding. The voltage of two turn step is then 138.56 × 2 = 277.12 V. This voltage produces 564.7A magnetizing current. The magnetizing power per phase is $S_{mag/\varnothing,\ PA}$ = 564.7 × 277.12/10^3 = 156.5 KVA. The calculated no-load loss is 3674 W.

The total magnetizing power is 14.7 + 1.3 + 156.5 = 172.5 kVA. The magnetizing current read from LV bushing is I_{mag}(A) = 172.5/(19.56/ $\sqrt{3}$) = 15.3 A; I_{mag} (%) = 15.3/1336 × 100 = 1.15%. Where 1336 A is LV rated current at tap 15 L. The total no-load loss is 39,149 + 4050 + 3674 = 46,873 W. The current due to no-load loss is I_{NLL}(%) = 46873/(50 × 10^6) × 100 = 0.094%. The exciting current is

$$I_{exciting} = \sqrt{1.15^2 + 0.094^2} = 1.15\%$$

The joint factor of 1.0 is used in these calculations; this means that the gaps between the yoke and leg are not considered. Even so, the calculated exciting currents are sometimes higher than the test values, which may indicate that the calculated core weights are higher than actual ones.

2.5 INRUSH CURRENT

When a transformer is switched off, the exciting current is down to zero. The flux, however, follows hysteresis loop reducing to a level but not zero; this flux is called residual flux, ϕ_r. When the transformer is switched back to the power network, a new flux, ϕ, is generated. Because flux cannot be changed at the moment of switching, i.e., $\phi_{t=-0} = \phi_{t=+0}$, when the new flux is not same as the residual flux, the flux has to change from the residual flux to the new flux, a flux-transient process occurs. The exiting current in the transient process, which is called inrush current, may be much higher than normal steady state exciting current. Also, inrush current has a different wave shape from normal exciting current. The inrush current phenomenon happens

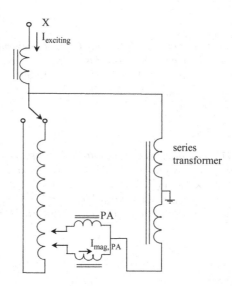

FIGURE 2.4 LV regulating winding connected to PA and series transformer.

only in primary winding; the secondary winding doesn't have such abnormal current during this period. Several cases are studied here.

(a) Switching in at zero voltage and no residual flux as shown in Figure 2.5a. At the instant of switching in, there is no residual flux in the core; the flux must start from zero. To maintain the first half cycle of the voltage wave, the flux density reaches a value corresponding approximately to twice that of the rated maximum flux density. The inrush current magnitude reaches a level possibly many times that of the normal exciting current and may exceed the full load current.

(b) Switch in at zero voltage while the residual flux having the maximum magnitude and opposite polarity to one produced by applied voltage. Instead of starting from zero, the flux has to start at the value corresponding to the polarity and magnitude of the residual flux in the core as shown in Figure 2.5b. In the first cycle the flux will reach about three times that of the rated maximum flux density, if there is no resistance in primary winding circuit, the inrush current is even greater.

(c) Switch in at zero voltage while the residual flux has maximum magnitude and the same polarity as one produced by applied voltage. The flux will follow its normal course and there is no inrush current, as shown in Figure 2.5c. Switching in at maximum voltage and no residual flux. From Equation (2.1), the time of the maximum voltage is the time of zero flux, so there is no transient process, the time of switch in is the time the flux is on the normal steady changing curve. As the result, there is no inrush current, as shown in Figure 2.5c.

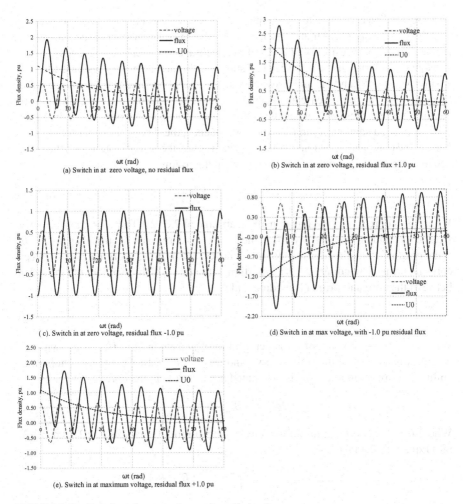

FIGURE 2.5   Relations between flux and applied voltage.

(d)   Switch in at maximum voltage while the residual flux having maximum magnitude and opposite polarity to one produced by applied voltage. In the steady state case when the voltage reaches its maximum, the flux in the core will be zero. Due to the residual flux in the core, at the time of switch in, the flux in the core is forced to be same as the residual flux. A transient process happens as shown in Figure 2.5d. As can be seen, the flux can reach two times the rated maximum flux; correspondingly, the inrush current is high.

(e)   Switching in at maximum voltage while the residual flux has maximum magnitude and polarity being the same as one produced by applied voltage. A transient process occurs, as shown in Figure 2.5e. This transient can cause nearly two times the rated flux density and a high inrush current.

Six cases studied earlier are for single-phase transformer operation. The principles apply equally well to poly-phase transformer, provided that the relationship between the phases is considered. The inrush current decays fast and disappears after a few seconds.

Now considering the worst scenario shown in Figure 2.5b. The residual flux has positive value, $+B_r$, and the unit is switched in when applied voltage is on its zero point. Since the flux cannot increase and decrease instantaneously, the flux starts from residual flux, $+B_r$, reaches $2\phi_{max}$ when the voltage reaches its positive peak and it drives the core far beyond saturation limit. As the consequence, the flux is no longer totally confined in the core, just a portion of it remains in the core, the rest of it spills into the space between the core and primary coil, as shown in the following equation:

$$2\phi_{max} = \left(B_s - B_r\right) \cdot A_c + B_{oil} \cdot A \tag{2.11}$$

Where B_s is core saturation flux density in arrange of 2.0~2.1 Tesla for grain-oriented steel, B_r is residual flux density left in the core, A_c is core cross-sectional area, $B_s - B_r$ is the portion of the flux density staying in the core; A is cross sectional area between core and primary winding; B_{oil} is the portion of the flux which is spilled into the space between the core and primary winding and is

$$B_{oil} = \mu_0 \cdot N \cdot I_{inrush} / H_{wdg} \tag{2.12}$$

Where I_{inrush} is peak value of inrush current in primary winding, H_{wdg} is the winding height, N is number of turn of primary winding, $\mu_0 = 4\pi \times 10^{-7}$ H/m, vacuum permeability. For approximation, ϕ_m is calculated as all of it is in the core

$$2\phi_m = 2 \cdot B_{max} \cdot A_c \tag{2.13}$$

Where B_{max} is rated maximum flux density. Substituting Equations (2.12) and (2.13) into Equation (2.11), then

$$N \cdot I_{inrush} = \frac{H_{wdg} \cdot A_C}{\mu_0 \cdot A}\left[2B_{max} - \left(B_s - B_r\right)\right] \tag{2.14}$$

Example 2.3

A step-down unit has a core with an area and diameter of 0.6055 m² and 0.914 m respectively. The mean diameter, height and number of turns of HV winding are 1.481 m, 2.101 m and 523 respectively. The rated maximum flux density is B = 1.68 Tesla. It is assumed that the residual flux is B_r = 1.68 × 0.9 = 1.51 Tesla. B_s = 2.1 Tesla is taken as core saturation flux density. The area between the core and HV winding is A = (1.481² − 0.914²)/4 × π = 1.067 m²
  The inrush current is

$$I_{inrush} = \frac{2.101 \times 0.6055}{4\pi \times 10^{-7} \times 1.067 \times 523}\left[2 \times 1.68 - \left(2.1 - 1.51\right)\right] = 5025\ A$$

$$I_{inrush}\% = 5025/702.9 \times 100 = 715\%$$

Where 702.9A is HV rated current.

Large inrush current impose a considerable mechanical stress on the transformer winding which needs to be checked, especially for large transformers. This current may cause protection relay to trip off, jeopardizing transformer insulation because interrupting the current with such magnitude may give rise to overvoltages exceeding the switching surges that normally happen in the network. Small transformers have a higher decay rate of inrush current. Transformers having higher losses have higher decay rates, too. When a transformer is switched in, an inrush current as high as 6~8 times the rated current flows in a transformer winding. Other transformers which are connected to the same network and are near the transformer being switched may also have a transient magnetizing current of appreciable magnitude at the same time.

2.6 TEST FAILURES OF NO-LOAD LOSS

Turn-to-turn insulation damage can be detected by higher no-load loss tested. A turn-to-turn failure happens if the turn-to-turn insulation is damaged. The damage may be the result of either poor production quality or dielectric test failure; for that reason, the no-load loss test is sometimes repeated after all of the dielectric tests are done. Turn-to-turn damage may not cause appreciable change in transferring function of its winding; hence there is no significant change in impulse wave form from the impulse test, it means that the impulse test may fail to detect turn-to-turn insulation failure. Figure 2.6 shows one example of turn-to-turn insulation damage which causes no-load loss tested much higher than what is calculated. The insulation papers of the winding cables were damaged during the production process.

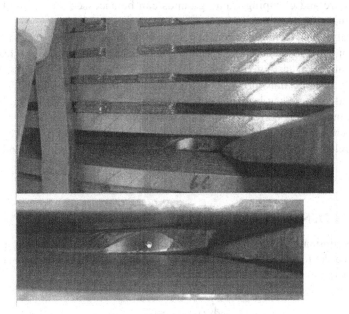

FIGURE 2.6   Photos of conductor paper damage causing high no-load loss.

Another type of failure of very high no-load loss is related to burrs at the lamination edge. As discussed earlier, when the burrs are long enough, they short out the laminations next to each other, making circulating current flow in laminations and producing extra loss.

2.7 CORE INSULATION AND GROUND

The core is insulated from its clamping structure and any other metallic parts, then it is grounded at one point, usually in the top yoke, by inserting one end of a copper wire or strip into the yoke, its another end is brought electrically outside the tank through a bushing and then grounded. Such structure ensures that the core has only one point grounded, rather than multiple grounding points. Multiple grounding points could make conducting loops, causing circulating current to flow and result in local heating and gassing. Thermal capability has to be considered when selecting the insulation material since the core surfaces that the insulation materials touch are sometimes of high temperature; a low thermal rating material could shorten the service life of core insulation.

Three facts need to be addressed when considering grounding [4,9]. First, any metallic parts in the transformer, unless solidly grounded, will get an electric potential in operation. The magnitudes of the potentials depend on their locations in the electric field. Placing the clamping frame directly on the tank base is a convenient way for a medium-size transformer. For a large important unit, the core clamping frame is isolated, only one ground point, like the core ground point, is brought through the tank via bushing and then grounded externally. This way the core and clamping frame grounds can be accessed during field service maintenance.

Secondly, when there is a conducting loop inside the tank, there may be a circulating current in the loop due to leakage flux. The circulating currents could create extra stray loss and local heating. Such loops in the tank, frame and core should be avoided.

Finally, in large transformers, third harmonic flux and stray 60 Hz flux from unbalanced voltage may leave the core and enter free space inside the transformer. This flux induces the currents in internal metallic parts of the transformer and may cause severe localized overheating. The five-leg core solves this problem by providing low reluctance paths around the three core legs between the top and bottom yokes.

2.8 FLUX DENSITY GENERATED BY QUASI-DC CURRENT

A very preliminary attempt is carried out here to get flux density under quasi-DC voltage in order to understand the effect of geomagnetically induced current on transformer operation. The ampere circuital law is

$$\oint \vec{H} \cdot \vec{dl} = NI \tag{2.15}$$

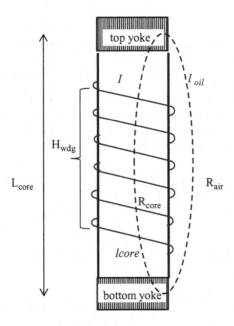

FIGURE 2.7   Sketch of a side view of a core and coil.

Where H is the magnet H-field, A/m; dl is an infinitesimal element of the path l shown in Figure 2.7, m; N is the number of turns of winding on one leg; I is quasi DC current in the winding, A. As an approximation, assuming H is same at anywhere along the path l and its direction is the same as l, then

$$H = \frac{NI}{l}; \ B_{DC} = \mu_0 \cdot H = \mu_0 \frac{NI}{l} \qquad (2.16)$$

Where B_{DC} is flux density, Tesla; $\mu_0 = 4\pi \times 10^{-7}$ H/m, vacuum permeability. The flux, ϕ_{DC}, is

$$\phi_{DC} = B_{DC} \cdot A_{DC} = \mu_0 \frac{NI}{l} \cdot A_{DC} = \frac{NI}{R_{m\,DC}} \qquad (2.17)$$

Where $R_{m\,DC}$ is reluctance, $R_{m\,DC} = l/(\mu_0 A_{DC})$, A_{DC} is the cross-sectional area which ϕ_{DC} goes through. On the flux path, there are two reluctances in series:

$$R_{m\,DC} = R_{core} + R_{air} = \frac{l_{core}}{\mu_0 \cdot \mu_r \cdot A_c} + \frac{l_{oil}}{\mu_0 \cdot A_{oil}} \qquad (2.18)$$

Where l_{core} is the length of the path in the core, m, l_{oil} is the length of the path in oil, m. A_c, A_{oil} are cross-sectional areas of the core and of oil path respectively, m^2. The discussion is for three-leg core three phase unit. For five-leg core three phase unit, the situation is totally different.

Example 2.4

A three-leg core three phase unit has HV winding wye connected, 528 turns, and its core has a cross-sectional area of 0.618 m², window height of 2.54 m and yoke height of 0.86 m. The tank width and length are 2.794 m and 5.944 m respectively. GIC at neutral is 200 A. It is assumed that when the flux comes out the core, it is confined in the tank, so $A_{oil} = 2.794 \times 5.944 - 3 \times 0.618 = 14.75$ m². The length of flux path in the core is assumed to be core leg height plus one yoke height: $l_{core} = 2.54 + 0.86 = 3.4$ m. The length in the oil is assumed to be the same as the length in core. The reluctance $R_{m\,DC}$ is

$$R_{m\,DC} = \frac{1}{4\pi \times 10^{-7}} \left(\frac{3.4}{4000 \times 0.618} + \frac{3.4}{14.75} \right) = 184527\,\Omega$$

The flux is

$$\phi_{DC} = \frac{NI}{R_{m\,DC}} = \frac{528 \times 200/3}{184527} = 0.191\,\text{Wb}$$

The flux density in the core is

$$B_{DC} = \frac{\phi_{DC}}{A_c} = \frac{0.191}{0.618} = 0.309\,\text{Tesla}$$

It should be noted that a few of assumptions are used in the estimation. Accurate calculation can be done by Finite Element Analysis.

2.9 GAPPED CORE

Reactive type on-load tap changers are often used to regulate LV or HV voltages. A preventive auto transformer (PA), or reactor, is needed for this tap changer application. The core of the reactor is of gapped core, i.e., there are several gaps in the leg, as shown in Figure 2.8 which has two gaps in each leg. In such a core, even though the iron portions have nonlinear relation of flux with current, the complete magnetic circuit, i.e., the core has a fairly linear relation due to high reluctance of the air gaps, it means that the reactance is fairly constant like air core. The reason why gapped core, and not air core, is selected is that the gapped core can take higher flux density and constrain most of them in the core. It makes the reactor overall size small and reduces stray flux and overheating it may cause in metallic parts. Because of high reluctance of the gaps, higher exciting current is required to drive flux through the core. Further the exciting current is more sinusoidal-like. The relation of the exciting current with the gap thickness and other factors is derived below. The maximum flux density, B_{max}, in core is shown in Equation (1.4) and repeated here:

$$B_{max} = \frac{V/N}{\sqrt{2}\pi fA_{Fe}} \qquad (2.19)$$

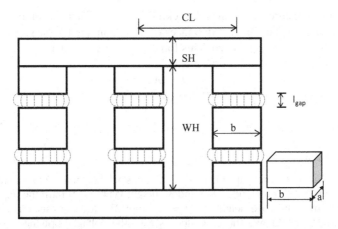

FIGURE 2.8 Gapped core.

Where N is number of turn in winding; A_{Fe} is the net core cross-sectional area, m²; V is voltage applied on the reactor's winding, Volt; f is frequency, Hz. The effective flux density, B_{eff}, in the core, then, is

$$B_{eff} = \frac{B_m}{\sqrt{2}} = \frac{V/N}{2\pi f A_{Fe}} \qquad (2.20)$$

The effective flux in the core is

$$\phi_{eff} = B_{eff} \cdot A_{Fe} = \frac{V/N}{2\pi f} \qquad (2.21)$$

The same flux goes through the gaps. The flux can also be presented as

$$\phi_{eff} = \frac{N \cdot I_{exciting}}{\left[\Sigma l_{gap} / \left(\mu_0 \cdot A_{gap} \right) \right]} = \frac{N \cdot I_{exciting}}{\Sigma l_{gap}} \cdot \mu_0 \cdot A_{gap} \qquad (2.22)$$

Where $I_{exciting}$ is exciting current, A, Σl_{gap} is total gap thickness in one leg, m, $\mu_0 = 4\pi \times 10^{-7}$ H/m, vacuum permeability, A_{gap} is cross-sectional area of gap, m². The exciting current is

$$I_{exciting} = \frac{V \cdot \Sigma l_{gap}}{2\pi \mu_0 \cdot f \cdot N^2 \cdot A_{gap}} \qquad (2.23)$$

Consider the flux fringing in the gap as shown in Figure 2.8. The equivalent cross-sectional area the flux goes through may be expressed as

$$A_{gap} = \left(a + 2l_{gap} \right) \left(b + 2l_{gap} \right) \qquad (2.24)$$

Where b and a are core leg width and stacking thickness. There is compressive force in the gap due to the fact that the two faces of iron bounding the air gap attract each other when the core is magnetized. This force, f_{gap}, is expressed as

$$f_{gap} = \frac{B_{eff}^2 \cdot A_{gap}}{8\pi\mu_0} \qquad (2.25)$$

Example 2.5

A reactor has 4 of 8 mm gaps in each leg, a core stack height $a = 235$ mm and the leg width $b = 182$ mm. The core window height WH $= 450$ mm, the leg center CL $= 425$ mm, the yoke height is SH $= 190$ mm. The leg cross-sectional area is $235 \times 182 = 0.043$ m². The core steel is M6. The voltage applying on coil is 173.205 V, 60 Hz. The coil has N $= 16$ turns. The volts per turn is V/N $= 173.205/16 = 10.825$. The flux density in the leg steel is

$$B_{eff} = \frac{V/N}{2\pi f A_{Fe}} = \frac{10.825}{2\pi \times 60 \times 0.043 \times 0.9} = 0.74 \text{ Tesla}$$

Where 0.9 is core stacking factor, the exciting current needed to drive the same amount of flux through the gaps is per Equation (2.23):

$$I_{exciting} = \frac{173.205 \times 4 \times 8 \times 10^{-3}}{2\pi \times 4\pi \times 10^{-7} \times 60 \times 16^2 \times 0.05} \approx 914 \text{ A}$$

Where $0.05 = (235 + 2 \times 8)(182 + 2 \times 8) \times 10^{-6}$ m², is the cross-sectional area of the gap. This current doesn't include the current needed to drive the flux through the core steel. From the B-H curve of steel M6, with 0.74 Tesla, the magnetizing force $H_s = 47$ A/m. The length of the flux going through is estimated as $l_s = (450 + 190 + 425 - 4 \times 8) \times 10^{-3} = 1.033$ m. The magnetomotive force in the steel path is $F_s = H_s \cdot l_s = 47 \times 1.033 \approx 49$ A. The flux density in the gap is

$$B_{gap} = B_{eff} \times \frac{0.043 \times 0.9}{0.05} = 0.74 \times \frac{0.0387}{0.05} = 0.57 \text{ Tesla}$$

The magnetizing force in the gap, H_{gap}, is

$$H_{gap} = \frac{B_{eff}}{\mu_0} = \frac{0.57}{4\pi \times 10^{-7}} \approx 4.56 \times 10^5 \text{ A/m}$$

The magnetomotive force in the gap is $F_{gap} = H_{gap} \cdot l_{gap} = 4.56 \times 10^5 \times 4 \times 8 \times 10^{-3} = 14592$ A. The total magnetomotive force is $F_{total} = F_s + F_{gap} = 49 + 14{,}592 = 14{,}641$ A. The exciting current is $I_{total\ exciting} = F_{total}/N = 14{,}641/16 \approx 915$ A. The exciting current required in the gap is about 99.9% (914/915) of the total exciting current.

REFERENCES

1. R. Feinberg, *Modern Power Transformer Practice*, John Wiley & Sons, New York, 1979.
2. Robert M. Del Vecchio, et al., *Transformer Design Principles with Applications to Core-Form Power transformers,* Second Edition, CRC Press/Taylor & Francis Group, Boca Raton, London, New York, 2010.
3. Barry W. Kennedy, *Energy Efficient Transformers,* McGraw-Hill, 1998.
4. Martin J Heathcote, *The J & P Transformer book*, 13th edition, Elsevier, Newnes, Amsterdam, et al., 2007.
5. John J. Winders, Jr., *Power Transformers, Principles and Applications,* Marcel Dekker Inc., New York, Basel, 2002.
6. Ramsis S. Girgis, et al., *Experimental Investigations on Effect of Core Production Attributes on Transformer Core Loss Performance, IEEE Trans. On Power Delivery*, Vol. 13, No. 2, April, 1998, 526–531.
7. K.O. Lin, *Transformer Stacked Core Structure*, Wound magnetics journal, January–March, 1995.
8. K. Karsai, et al., *Large Power Transformers*, Elsevier Science Publishing Company Inc., Amsterdam, Oxford, New York, Tokyo, 1987.
9. S. V. Kulkarni, et al., *Transformer Engineering, Design and Practice*, Marcel Dekker, Inc., New York, Basel, 2004.
10. L. F. Blume, et al., *Transformer Engineering*, Second Edition, John Wiley & Sons, Inc., New York, Chapman & Hall, Ltd., London, 1951.

3 Windings

Windings are perhaps the most important component in power transformers. Before discussing their performances such as insulation, short circuit and thermal, some of the characteristics of windings, such as types and application of windings, transpositions and half turn, are discussed here.

3.1 TYPES OF WINDING

3.1.1 LAYER WINDING

Conductor cables of a layer winding are wound in axial direction in helix fashion without radial spacers between turns and cables, as shown in Figure 3.1. Because there is no radial spacer, turn-to-turn insulation and cooling performances of the layer winding are weak compared to other types of windings, and also when the winding radial build (winding outer radius–winding inner radius) is thin, for example less than 10 mm, its mechanical strengths against short-circuit forces are marginal. On the other hand, it costs less in labor and material. This type of winding is usually used when both rated voltage and rated current are low. It is also one type of LV regulating (tap) windings that are commonly used.

The cables used are single wires. In most cases, the cables are wound on the flat side. In some cases, the cables are wound on the edge in order to reduce the eddy loss caused by radial leakage flux hitting on the winding ends as a measure to reduce the winding hot spot temperature. In this case, the thickness of the cables should be sufficient compared to its width, so that the cables remain twist free in production [1].

As for regulating winding, there are two ways to make tap leads. One way is to braze the tap leads at designate turns and bring them to the top of the winding, as shown in Figure 3.1. This winding has two portions, top and bottom, which mirror each other to get ampere-turns balanced along the winding axial direction. However, along the radial direction, the ampere-turn balance with adjacent winding is worse since only a fraction of the winding carries current except at extreme taps; this makes the winding mechanically weak. Another way to make tap leads is to make multi-start winding, discussed in the next subsection. With multi-start style, several layers can be made by winding one layer on top of another. Most practical designs have either one or two layers, and one advantage of two-layer winding is that all leads are brought to the top, making it easy for routing lead cables. More importantly, the current direction of start leads is opposite that of the current direction of finish leads, the sum of current of all leads is zero, in such a way that no magnetic field is generated by the leads as the result and no local overheating in metallic components nearby, such as the core clamping frame, occurs.

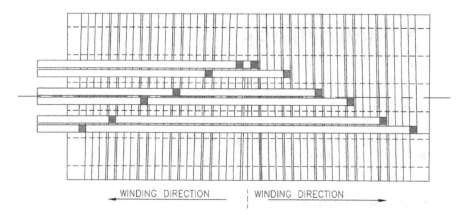

WINDING DIRECTION   WINDING DIRECTION

FIGURE 3.1   Layer tap winding.

3.1.2 MULTI-START WINDING [2,3]

Multi-start winding, as shown diagrammatically in Figure 3.2, is basically several helical windings wound together with or without radial spacer. It is used as regulating winding. The advantage of this type of winding is that each tap has its turns distributed uniformly along the whole length of winding so the ampere-turn and magnetic

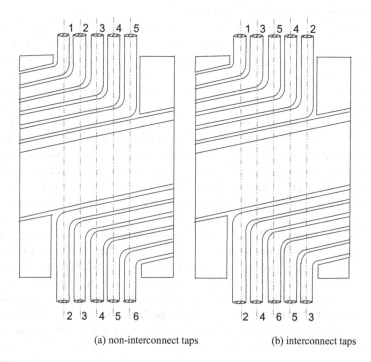

<table>
<tr><td>(a) non-interconnect taps</td><td>(b) interconnect taps</td></tr>
</table>

FIGURE 3.2   Multi-start winding.

balances are achieved, which significantly reduce short-circuit forces. Secondly, bigger cable such as CTC (continuously transposed cable) can be used for higher-current units. The following points need to be considered when selecting multi-start winding.

First, the electric stresses between taps and between turns have to be considered carefully. As shown in Figure 3.2a, if the tap lead is numbered in sequence order, the voltage between the first turn of tap 5 and the second turn of tap 1 is about four times that of the tap voltage. This arrangement also has low series capacitance; as a consequence, high oscillation during impulse test is expected. The way to mitigate the voltage between taps is to interconnect the taps, as shown in Figure 3.2b. Such lead arrangement restricts the operating voltage between two adjacent lead cables to no more than twice that of the tap voltage. This arrangement also has a high series capacitance, which improves the voltage distributions under transient voltages. The multi-start winding can be used as regulating (tap) winding in various tap arrangements, such as liner, coarse/fine and plus/minus taps, which are diagrammatically shown in Figure 3.3.

The second concern is its mechanical strength. The multi-start winding is basically a helical winding. Under short-circuit conditions, if it is outmost winding which experiences an outward tension force, with a small number of turns wound in helical fashion offering little resistance to this outward bursting force, the winding's ends must be very securely restrained to ensure that the winding does not unwind itself under such force; the pressboard sticks placed around the outer circumference of windings keyed into radial spacers help to some extent. The most common improved alternative to this, when the tap winding is outmost winding, is to use disc tap winding.

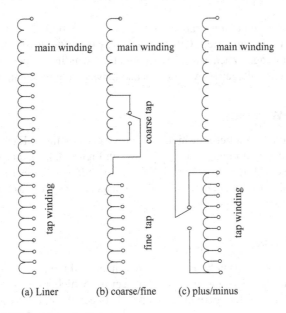

FIGURE 3.3 Different tap windings.

The number of turns between taps is the number of turns of each individual helical coil in the multi-start winding. A minimum of two turns are needed to make the individual helical coil a closed coil, in such a way as to keep the coil mechanically tight and strong. If the individual helical coil has only one turn, the coil is open and loose, and it is hard to tighten it up and to restrain its leads, which makes the whole winding mechanically weak, so it is not good design and practice.

The third concern is that the currents in all the leads at either the top or bottom have the same direction. The magnetic field generated by the sum of these currents is sometimes so high that it could generate significant eddy current in adjacent metallic parts, producing extra loss and local heating in these parts, such as the core clamping frame, tank walls and cover, to unacceptable levels. One preventive measure is to use two tap windings, the directions of which are opposite each other, to make the direction of the current in each winding opposite each other. As the result, the sum of lead currents is zero, and the flux generated is greatly reduced so as to minimize the loss and the local heating in the metallic parts. This practice is recommended when the sum of currents is over 5 KA. However, when the tap winding is outermost winding, which is far from the clamping frame, the current could be higher.

Finally, compared to layer regulating winding having brazed tap leads, multi-start windings require more labor and time to make.

In large transformers, a separate regulating winding for HV taps is normally used when gap between taps in main windings upset the electromagnetic balance to such an unacceptable degree that the short-circuit forces exerted on the winding in the event of an external fault is too high for the winding to withstand. It should be noticed that the separate regulating winding costs significantly more because additional winding, additional insulation gaps between windings and larger tank sizes are required. Therefore, it is always preferable to accommodate the taps in the main body if possible.

When very large regulating ranges, say ±20%, are required, a two-regulating-windings structure is suggested. Otherwise, it will be difficult to handle creepage stress in the winding. When using two regulating windings, special attention is needed on transient voltages between loops and between the two windings.

3.1.3 HELICAL WINDING

The turns are wound in a helix fashion on a pressboard cylinder. Each turn is separated from the next by a radial spacer, as shown in Figure 3.4. This type of winding is usually selected as low voltage and high current winding, one electrical turn could consist of a number of conductors. Helical windings are easy to make and mechanically strong.

3.1.4 DISC WINDING

This type of winding is usually used as high voltage and medium current winding. The winding consists of a number of discs or sections which are connected in series and placed in an axial direction with an oil duct between them. Each disc consists of a number of turns wound in a radial direction, as shown in Figure 3.5. Each electrical

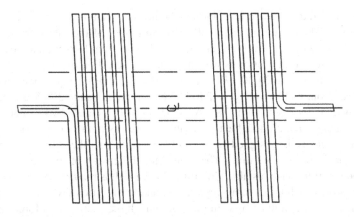

FIGURE 3.4 Helical winding.

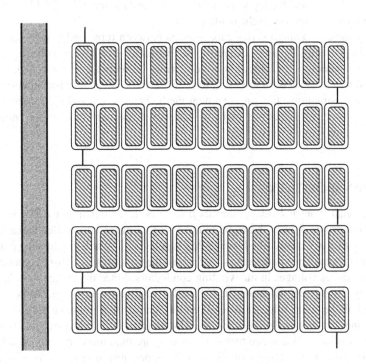

FIGURE 3.5 Disc winding.

turn may consist of several conductors based on current magnitude and cooling requirements. In the case of more than one conductor needed to make one turn, transposition has to be made to reduce eddy current loss in the conductors.

For wye-connected HV winding having graded insulation, which means its neutral is grounded, there is an option to make the HV line lead at the center of the winding; the top half winding and the bottom half winding are parallel connected, and the leads from top and bottom of the winding form the neutral lead. The advantage of this structure is that the requested insulation clearance between the ends of winding and yoke is small. The current in each half winding is half of the total current, which greatly reduces short-circuit forces applied on the winding. However, not every HV winding with graded insulation is well designed by selection of such structure. For example, a unit has 18/24/30 *MVA* ratings, HV winding is wye-connected with graded insulation, its line voltage is 230 kV at nominal tap and 220 kV at minimum tap, and the maximum current is 78.7 A. If HV winding is made by two parallel circuits as previously mentioned, the maximum current in winding conductor is 39.4 A. The current density is designed to be about 3 A/mm^2, so the required conductor cross-sectional area is 13.1 mm^2. The conductor dimension of this cross-sectional area is too small to make a stable winding. This case thus requires either increasing conductor size, which needs more copper, or making one circuit winding. The study showed that one circuit winding is more economical; the short-circuit forces are still small because the current is small. The top lead is the line lead, the clearance between the top of winding and the top yoke has to be increased to meet the 230 kV insulation class, and the bottom lead is the neutral lead.

Disc winding has high series capacitances between turns and between sections, giving the winding a better dielectric performance. For higher voltage, floating shielding conductors are placed in several pairs of sections near the high voltage lead, or interleaved sections are used, to improve the voltage distribution under transient voltage situations; this improvement method can be utilized only in disc winding. The structure of sections make disc winding mechanically strong enough to withstand short-circuit forces.

3.2 TRANSPOSITIONS

When a transformer is loaded, currents flow in both LV winding and HV winding, and these currents produce a magnetic field the flux of which is called leakage flux. Unlike the flux in the core, the leakage flux is generated by winding loading currents, and it exists outside the core in the space where the windings are. This magnetic leakage flux induces voltages in the winding conductors. Since the magnetic field is not uniformly distributed, as shown in Figure 1.4 in Chapter 1, the magnitude of the induced voltage in a conductor depends on its location; that is, the nearer to the main flux channel (the barrier between LV and HV windings) the conductor is, the higher its induced voltage. When one turn consists of more than one conductor, conductive loops exist, due to the crimps of all conductors per turn in order to form start and finish leads, as diagrammatically shown in Figure 3.6a. In helical and disc windings, the conductors of one turn may be placed side by side radially sometimes, such an arrangement makes each conductor have a different location in the leakage magnetic field, and consequently, the voltage induced in each conductor is different. The sum of these voltages in the loop shown in Figure 3.6a is then not zero, and this non-zero voltage produces circulating current in the loop which generates so-called circulating current losses. In some large units designed improperly, the circulating currents are so high that they cause higher-than-expected load losses, overheat the conductor insulation papers and eventually bring failures.

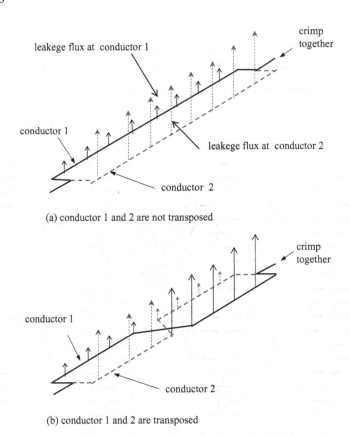

(a) conductor 1 and 2 are not transposed

(b) conductor 1 and 2 are transposed

FIGURE 3.6   Leakage flux in conductors 1 and 2 which make one turn.

The circulating current can be reduced by reducing the sum of induced voltages in the loop; this is the purpose of transposition. The transpositions make each conductor have the same spatial location to get the same induced voltage, as shown in Figure 3.6b, to make the sum of induced voltage in a loop and the circulating current losses as small as possible.

3.2.1 HELICAL WINDING

When one turn consists of n cables which are wound side by side in a radial direction, ideally, $n - 1$ transpositions are needed to make each cable occupy every location equally. Figure 3.7 shows a transposition of three cables. With continuously transposed cable, $n - 1$ transpositions can be realized since few than ten cables are normally used. For multi-strand helical winding, more cable wires are used to make one turn; for example, when 20 wires are selected to make one turn, 19 transpositions are needed ideally, in practice, which could cause a lot of work. In such cases, only a few transpositions (usually three) are made, as shown in Figure 3.8, to reduce the circulating current loss to a tolerable level.

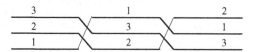

FIGURE 3.7   Transpositions of three radially placed conductors.

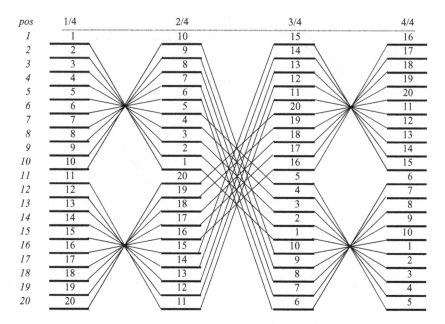

FIGURE 3.8   Multi-strand helical winding transpositions.

The preceding discussions focus on the losses generated by axial leakage flux. In the cases when losses are generated mainly by radial leakage flux, so-called rolling transposition, as shown in Figure 3.9, is used to reduce these losses by positioning each wire at every axial position equally.

3.2.2 Disc Winding

When one turn consists of one wire or one cable, transposition is not needed. When one turn consists of more than one cable, for example two cables, transposition shown in Figure 3.10a is commonly used. The interleaved winding and shielded winding are also shown in Figure 3.10b and c; their applications are not for loss reduction but for improvement on the winding dielectric strength under impulse voltages.

3.3 HALF-TURN EFFECT

It is sometimes desirable to have start lead and finish lead of a winding brought out on opposite sides of the coil, such as series winding of autotransformer, for example. In order to achieve it, a half physical turn is used, as shown in Figure 3.11. The

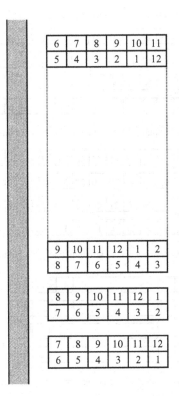

FIGURE 3.9 Rolling transposition.

half physical turn in each winding could cause one electrical turn difference between the side legs, depending on whether this half turn goes through the core window. This turn difference causes extra flux. In the three-leg, three-phase unit, this extra flux has to leave one yoke, go through the air/oil path to enter into another yoke. Due to high reluctance of the air/oil path, the flux is reduced to negligible level, so no extra loss is added. For the center leg, the electrical turns are nearly same as the physical turns. In a single-phase, three-leg transformer, the side legs and end yokes offer a low reluctance path to the flux made by the half turn. As the result, extra core loss is expected when the transformer has an unbalanced load. With same number of physical turns, N (integer + 0.5), the electrical turns on each leg are listed in Table 3.1.

Facing the side on which the winding starts, when the start leads go to the right, the winding on the right leg has 0.5 electrical turn less than the physical turn, while the winding on the left leg has 0.5 electrical turn more than the physical turn. A transformer with a larger number of physical turns, using the same number of physical turns on each leg in design calculations, couldn't cause a significant difference between real and designed performances even the electrical turns on each leg are different by one turn. On the other hand, with a small number of physical turns, the difference between electrical turns has impact on the performances.

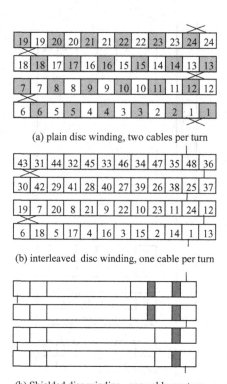

(a) plain disc winding, two cables per turn

(b) interleaved disc winding, one cable per turn

(b) Shielded disc winding, one cable per turn

FIGURE 3.10   Disc windings.

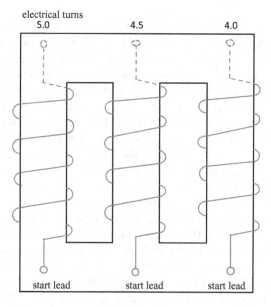

FIGURE 3.11   Half turn.

TABLE 3.1
Effective Electrical Turns

Core Type	Leg R	Leg M	Leg L
Single-phase, two-leg	$N - 0.5$		$N + 0.5$
Three-phase, three-leg	$N - 0.5$		$N + 0.5$

3.4 AXIAL SPLIT WINDINGS

Some transformers have one set of HV winding connecting to one network, and two sets of LV windings, each of which is connected to a different network; these two sets of LV windings work independently. The two sets of LV windings can be sometimes be stacked one on top of another; this is called axial split winding design. HV winding is also split into top and bottom halves to balance magnetically the two sets of LV windings. Another option is to radially separate the two sets of LV windings; this is called radial split winding design.

The advantage of the axial split winding design over the radial split winding design is that there is a considerable saving on HV winding material and space inside the tank; as a result, the size of the whole unit is smaller. Figure 1.7 in Chapter 1 shows the sketch of the axial split and radial split windings arrangement. In the case of axial split winding, when one set of LV windings, for example LV1, has current, the corresponding portion of HV winding, HV1, also has current flow. When LV2 windings have no current, HV2, has current of 3~5% of HV1 current flow even it is electrically parallel connected with HV1; this is because the impedance between the HV2 and LV1 is much higher.

With same magnitudes of short-circuit current, the forces are higher in the case of only one set of LV windings being short-circuited, compared to both sets of LV windings being short-circuited. One concern or checking point is circulating current in HV winding, the circuit of which is made by two parallel circuits, top half and bottom half windings. When all same numbered tap leads are connected together, conductive loops are formed if one tap changer is used, as shown in Figure 3.12a. In the case of only one of the LV windings, for example the top Y winding, being loaded, the top half of HV winding is loaded, too. The leakage flux shown in Figure 3.13b induces a voltage, for example between tap 6 and tap 4. This induced voltage produces a circulating current in the loop consisted of tap 4 and tap 6 of top and bottom. Since the resistance of this loop is very small, the circulating current is so high that in some cases, it chars the winding and results in a failure. This type of failure is not easily found since the factory heat run test is conducted with both sets of LV windings loaded as shown in Figure 3.13c. In this situation, the induced voltages at tap 6 and tap 4 in the top half and bottom half are quite balanced, resulting in a very small circulating current.

To reduce the circulating current to a tolerable level, the following measures can be considered. First is to use the separate tap changers for the top and bottom halves to avoid the loops made by one tap changer, as shown in Figure 3.12b. The second

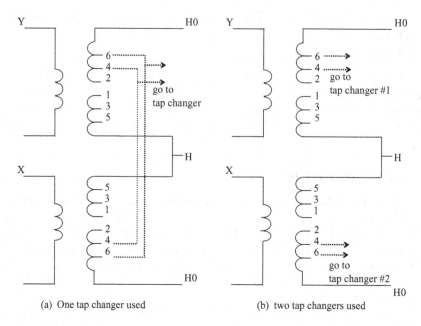

(a) One tap changer used   (b) two tap changers used

FIGURE 3.12 Circuit of axial split windings.

measure is avoiding placement of the tap sections near the winding ends. The radial flux component increases at the winding ends, producing higher induced voltage. Third is making an interleaved tap section to increase the resistance between taps. Some designs have regulating winding located inside its LV winding and have loops in the regulating winding due to other design considerations. When one of the LV windings, for example LV1, is loaded as shown in Figure 3.14a, its radial flux hits the ends of its own regulating winding RV1 and the top end of RV2, LV2's regulating winding, inducing a voltage which generates circulating currents in the loops. The circulating currents produced are sometimes large enough to cause overheating, resulting in damage to the insulations of winding cable and causing a failure. In order to reduce this circulating current, the regulating winding height can be shortened, so less radial flux hits on the windings ends.

3.5 CABLES USED IN WINDING

Wire cables such as single strand, twin strand or triple-strand cables as shown in Figure 3.15a, b and c are commonly used when the winding's current is not high. For high current winding, CTC (continuously transposed cable), as shown in Figure 3.15d, is used. The advantages of CTC come from its thin and small strands, which are transposed to make each strand occupy equally every spatial location within the CTC; this greatly reduces the eddy current losses in windings. The second merit is the higher amount of copper inside core window; it improves the space factor, which means that the core window area is effectively used and the unit's overall size could be reduced. It is said that a minimum circulating current loss is achieved by full transposition in one turn.

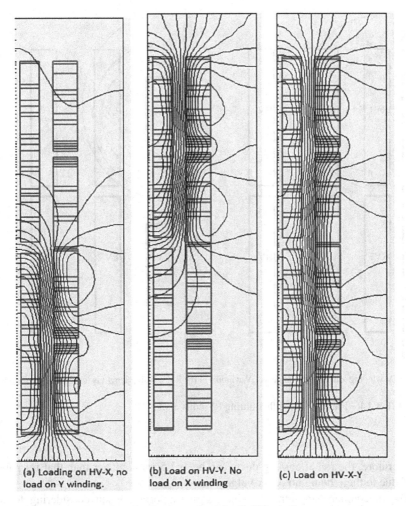

(a) Loading on HV-X, no load on Y winding.

(b) Load on HV-Y. No load on X winding

(c) Load on HV-X-Y

FIGURE 3.13 Leakage flux distributions of axial split LV windings.

Thermal upgraded electrical insulating Kraft papers are commonly used as paper insulation of the cables. The papers are anisotropic material with different mechanical properties in both machine direction and cross direction, which correspond to the strip width. During production, the cellulose fibers are oriented in the machine direction, bringing a higher tensile strength and stiffness. In the cases that higher winding temperatures are required, such as mobile transformers, Nomex insulation paper is commonly used.

The wrap should be tight enough to avoid a bulge which could otherwise block the oil duct to cause higher winding gradients. When cooling of winding is the dominant factor, such as LV windings of large units, paperless CTC, such as polyester net or perforated wrappings, is used. Such CTC can bring the winding gradients down.

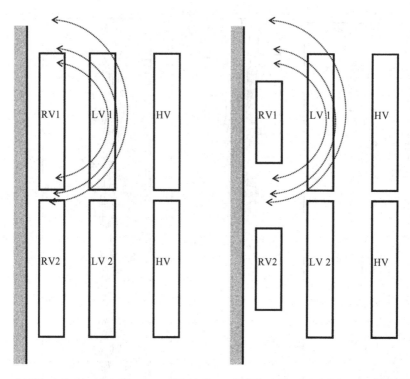

(a) RV winding length is the same as LV winding   (b) RV winding length is less than LV winding

FIGURE 3.14   Radial leakage flux hitting RV coils ends.

Furthermore, the radial cooling ducts can be 1~1.5 mm smaller than that for paper CTC due to there being no paper bulge issue.

The mechanical strengths of winding are important when considering forces exerted at the windings during short-circuit, high-yield strength of the copper can make winding strong to resist hoop and bending stresses. The yield strength can be increased by cold work which is a rolling process of the conductors. However, the cold work reduces the copper's conductivity as it elongates the grains in the copper. A reduction of the conductivity from 1% to 5% is observed [4]. To increase CTC bending stiffness significantly, a thermo-setting epoxy resin coat is introduced at the top of the enamel coat. When the winding is heated up to 120°C, the epoxy resin is cured which bonds all strands together behaving like one bar of the same overall size. The yield strength of the epoxy-coated CTC depends not only on copper yield strength but also on the shear strength of the epoxy coating. The shear strength of the epoxy coat can be tested based on ASTM D1002-10 [5]. The shear strengths not only at room temperature but also at as high a temperature as 120°C have to be considered since the winding temperature under full load is higher than room temperature. The maximum 50% reduction at 120°C compared to one at room temperature is usually accepted in the industry.

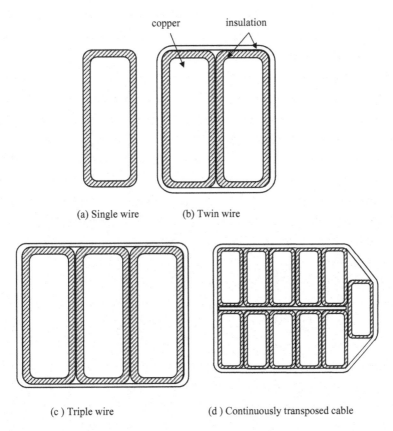

(a) Single wire   (b) Twin wire

(c) Triple wire   (d) Continuously transposed cable

FIGURE 3.15   Conductor cables.

REFERENCES

1. Bharat Heavy Electricals Limited, *Transformers*, McGraw-Hill, New York, Chicago, San Francisco, et al, 2003.
2. Martin J. Heathcote, *The J & P Transformer Book*, 13th edition, Elsevier, Newnes, Amsterdam, et al, 2007.
3. Jim Fyvie, *Design Aspects of Power Transformers*, Arima Publishing, London, 2009.
4. Deniel H. Geiβler, *Short-Circuit Withstand Capacity of Power Transformers*, Cuvillier Verlag, Göttingen, 2016.
5. ASTM D1002-10:2010, *Standard Test Method for Apparent Shear Strength of Single-Lap-Joint Adhesively Bonded Metal Specimens by Tension Loading*, 2010.

4 Insulation

The investigation on field failures shows that about 43% of failures originated from winding insulation, 19% from bushing, 16% from load tap changer [1]. A robust insulation structure is essential for trouble-free and long service life of a transformer. A good insulation structure makes a maximal use of intrinsic properties of insulation materials, offers effective measures in limited space to prevent from dielectric failures in oil, in solid insulation and at interfaces between them. It also can increase rated voltage and MVA rating without much increase in transformer physical size, so the economical design relies heavily on a well-designed insulation structure. This chapter provides detailed and useful information to assist readers in understanding and conducting good designs.

4.1 VOLTAGES ON TRANSFORMER TERMINALS

4.1.1 Service Voltage

Generally, service voltage is the rated voltage of the transformer. The system voltage always fluctuates; its highest level is maximum system voltage, and the insulation is designed to enable the transformer to work well under such a level of voltage. The maximum system voltage is slightly different from the rated voltage of the transformer; in most cases it is less than 5%.

4.1.2 Overvoltages

A transformer in a power system experiences four types of overvoltage. One type is caused by upsetting the symmetry of three-phase voltage, for example, single phase-to-ground fault for several hours. The second is caused by a sudden change of system, such as disconnecting a load at the remote end of a long transmission line; this overvoltage is maintained for a few seconds. The third one is caused by lightning in the atmosphere. The fourth one is caused by ferroresonance. Each overvoltage type is discussed in detail in this section.

4.1.2.1 Upset of Symmetry of Voltage

When single phase-to-ground short circuit happens, the voltage of the faulty phase drops to zero. At the same time, the voltages of sound phases may change depending on whether the neutral point is shifted. The magnitude of the neutral shift is a function of zero sequence impedance and positive sequence impedance of the system as viewed from the fault [2]. Its magnitude in the percentage of phase voltage is

$$Neutral\ shift\,(\%) = \frac{Z_0 - Z_+}{Z_0 + 2Z_+} \times 100 = \frac{k-1}{k+2} \times 100 \tag{4.1}$$

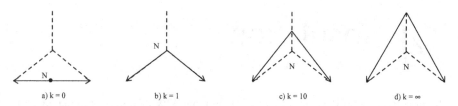

FIGURE 4.1 Vector graphics of sound phase voltages when 3rd phase is short-circuited.

Where $k = Z_0/Z_+$, and Z_0, Z_+ are zero sequence impedance, positive sequence impedance of the system at the point of fault respectively. As can be seen, different k values result in different shift magnitudes. The neutral shift results in the change of the voltage across the sound phase, as shown in Figure 4.1.

Four cases with system grounded through inductance are discussed.

(a) When zero sequence impedance is negligible compared to the positive sequence impedance, i.e., $k = 0$, the neutral shift is 50% negative (negative neutral shift is the neutral away from the faulty phase, positive neutral shift is towards the faulty phase), the voltages across sound phases are 86.6% of normal voltage, as shown in Figure 4.1a.

(b) When zero sequence impedance is equal to positive sequence impedance, $k = 1$, the neutral shift is zero, as shown in Figure 4.1b. One line fault doesn't affect the voltage across the sound phases. This is the case of delta-wye with delta primary and wye secondary, and also when the neutral of wye-connected windings is solidly grounded. There is no overvoltage in these cases.

(c) When $k = 10$, the voltage across the sound phases becomes about 1.5 times their normal voltage, as shown in Figure 4.1c. It is the case with wye-wye-delta tertiary with a small capacity of its tertiary. This situation occurs also on a multi-grounded distribution system.

(d) When $k =$ infinite, the neutral shifts completely; the voltage across sound phases becomes 173% or equal to the line-to-line voltage, as shown in Figure 4.1d. Under 105% voltage situation, which is allowed per IEEE C57.12.00, the sound phase voltage is 173% × 1.05 = 182%. In this case the neutral is ungrounded. or it is grounded through very high impedance.

It shows that from case (c) or (d), grounding through an impedance can reduce short-circuit current; as a result of this, the phase voltage of the sound phase increases. In some cases, double phase-to-ground faults cause overvoltage on sound phase slightly higher than single phase-to-ground, but single phase-to-ground fault occurs more frequently. If the dielectric strength of phase insulation is marginally over the normal operation voltage, the single phase-to-ground fault may come to grief on the transformer, and the system, since this voltage has service frequency, lasts long enough until the short-circuit fault is eliminated.

When neutral is grounded through a capacitor, or when a tertiary delta winding is closed through a capacitor, the net zero sequence impedance is negative or capacitive, and resonance occurs when this negative zero sequence reactance is twice the three-phase reactance. When the system is grounded through resistance, R_0, the zero sequence impedance is resistive, the positive sequence impedance is reactive and the ratio k is $-jk = R_0/jX_+$, then Equation (4.1) becomes

$$Neutral\ shift(\%) = \frac{jk+1}{jk-2} \times 100 = 25 + 75 \times \frac{jk+2}{jk-2} \tag{4.2}$$

It can be seen that in general, resistance impedance gives a higher phase voltage than inductive impedance due to phase shift [2].

4.1.2.2 Lightning Impulse

Lightning from the atmosphere strikes on a transformer. In actual power system operation, the transformer is sometimes subjected to power frequency voltage and lightning impulse voltage at the same time. The breakdown voltage under such a condition may be lower than the corresponding value for impulse alone [3]. The lightning impulse wave contains high-frequency components in its front (up to several hundred thousand Hertz), and low-frequency components in its tail. When such a wave enters the windings, the winding capacitance network reacts with its front at first, which mainly determines the voltage initial distribution along the winding. The winding inductance network determines the voltage final uniform distribution; here, the winding resistance network is not considered. The transition from the initial capacitive distribution to final inductive distribution involves oscillation. This oscillation is damped by core loss and winding resistance; however, in most cases, the maximum voltage happens in the first one or two oscillations, and the damping is not effective.

4.1.2.3 Switching Impulse

Switching actions, such as disconnecting a load at the remote end of a long transmission line, or switching in a long transmission line, cause so-called switching surge on the transformer. It lasts 200~2000 μs. Its frequency extends into the range of several thousand Hertz. It is highly damped, with magnitude of 2.2~2.5 times the peak value of rated phase-to-neutral voltage in most cases.

Like any network capable of oscillation, transformer windings consist of inductances and capacitances. When a network receives an impulse of single energy burst, in return a free oscillation may occur. When switching transients consists of an initial peak voltage followed by an oscillatory component the frequency of which coincides with a natural frequency of the winding, so-called part-winding resonance develops; its maximum magnitude depends on the damping of both external and the windings themselves, but it can occasionally be greater than the voltage resulting from the lightning impulse. Unlike the lightning impulse, where the transformer can be designed to withstand impulse voltage, the solution to the resonance problem cannot be achieved by the transformer manufacturer's action alone. Resonance requires a passive structure, which is transformer windings, and an active component represented by various sources of oscillating voltages which come from lightning, faults and switching. The detailed description can be found in reference [4].

Compared with power frequency voltages, the impulse waves have much shorter duration but much higher magnitude.

Ferroresonance

When a transformer core operates near saturation, the B-H curves are highly nonlinear, and the effective permeability of the core varies with the changes of flux density. This means that the inductance of the transformer changes with the flux. If the connecting cable runs fairly long, a significant amount of phase-to-ground capacitance may exist. Under this condition, a serious resonance almost certainly occurs at least part of the time during every cycle [5]. The condition for ferroresonance can be disputed by delta-connected secondary winding of the station service transformer if its primary is Y-connected. The delta-connected winding assures that the vector sum of the voltage of all three phases add to zero, stabilizing the neutral point of the Y-connected primary winding and preventing excessive voltage across the windings.

4.2 VOLTAGE INSIDE TRANSFORMER

4.2.1 ANALYSIS ON IDEAL MODEL

When a lightning or switching impulse hits a bushing, it also hits winding connected to the bushing. Due to high-frequencies content in the front of these impulses, the capacitances between turns, sections and windings, as well as the capacitances between windings and grounded parts, play a key role in initial voltage distribution along windings. Considering these capacitances and winding inductances, ignore the winding resistances and assume, first, the winding has same per-unit values (capacitance and inductance) along the winding length, as shown in Figure 4.2. Secondly, the impulse voltage is simulated by a step voltage to simplify the mathematical deduction; the change of voltage along the winding, $v(x)$, is per Kirchhoff's law

$$\frac{\partial^2 v}{\partial x^2} - LC\frac{\partial^2 v}{\partial t^2} + LK\frac{\partial^4 v}{\partial x^2 \cdot \partial t^2} = 0 \tag{4.3}$$

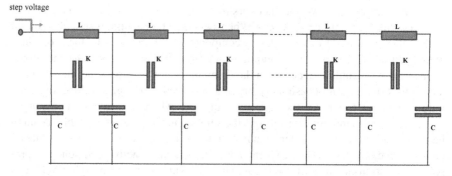

FIGURE 4.2 Simplified impulse calculation model of winding.

Where L is inductance per unit length, H/m, C is shunt capacitance per unit length, F/m, and K is series capacitance per unit length, F·m. The calculations of C, K and L are shown in reference [3]. At the moment of one end of winding being impulsed ($t = 0$), $L \to \infty$, meaning that the initial voltage distribution at $t = 0$ is dependent only on the shunt and series capacitances. Equation (4.3) then changes to

$$\frac{\partial^2 v}{\partial x^2} = \frac{C}{K} \cdot v(x) \tag{4.4}$$

When another end of the winding is grounded, the boundary conditions of Equation (4.4) is $v(0,0)=V$, and $v(l, 0)=0$. The solution to Equation (4.4) is

$$v\left(\frac{x}{l}\right) = V \cdot \frac{\sinh\left[\alpha\left(1 - \frac{x}{l}\right)\right]}{\sinh(\alpha)} \tag{4.5}$$

When another end of the winding is un-grounded, the solution is

$$v\left(\frac{x}{l}\right) = V \cdot \frac{\cosh\left[\alpha\left(1 - \frac{x}{l}\right)\right]}{\cosh(\alpha)} \tag{4.6}$$

where

$$\alpha = l \cdot \sqrt{C/K} = \sqrt{C_r / K_r} \tag{4.7}$$

where l is the winding length, $C_r = C \times l$ is the resultant shunt capacitance and $K_r = K/l$ is the resultant series capacitance. The initial voltage distributions along winding at $t = 0$ with different values of α are shown in Figure 4.3a for another end grounded, and Figure 4.3b for another end open. As can be seen, the higher the shunt capacitances are, the higher gradient of voltage drop on the first few sections. On the other hand, if one wants to improve this voltage distribution to be more uniform, greater series capacitance should be added, i.e. reducing α value.

The maximum voltage gradient, Δu_{max}, i.e. the voltage drop on the impulsed section is

$$\Delta u_{\max} = -\frac{\alpha}{l} \coth \alpha \tag{4.8}$$

The final distribution will be inductance dependent. For another end being grounded, the final distribution is

$$v\left(\frac{x}{l}\right) = V\left(1 - \frac{x}{l}\right) \tag{4.9}$$

For another end being open, the final distribution is

$$v\left(\frac{x}{l}\right) = V \tag{4.10}$$

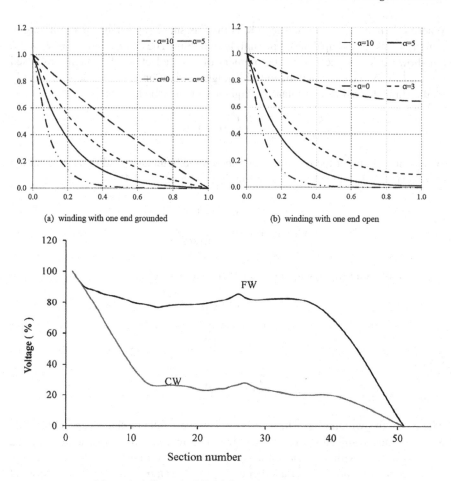

(a) winding with one end grounded

(b) winding with one end open

(c) a real winding under full and chopped waves with one end grounded

FIGURE 4.3   Voltage distribution in winding under impulse.

As the impulse wave travels further into the winding, the initial voltage distribution is modified due to the progressive effect of individual winding elements such as capacitance, self and mutual inductances and resistance. The traveling wave is reflected from the opposite end of the winding back towards the line end, and so on. These reflections interact with incoming wave, and a complex series of oscillation may occur and reoccur until the energy is dissipated [4].

From the capacitive initial distribution to the final inductive distribution, oscillations happen in winding. The amplitude of the oscillation depends on the value of α and damping factors.

Higher oscillation happens when the α value is high. Thus, small α value reduces not only non-uniformity of the initial capacitive voltage distribution, but also the amplitude of voltage oscillation. With certain insulation levels, winding clearance to the ground is fixed, which means that the shunt capacitances are almost unchanged. In this case, increasing the series capacitance to reduce the α value can depress the oscillation to some extent. In practice, three methods are used to increase series capacitance. One method is to place a static ring at the winding top and bottom ends. The static ring gives additional series capacitance to the section next to it and also mitigates the winding corner electric stress. A second method is to place floating wire in sections which are next to the impulsed terminal lead. The floating wires give additional series capacitance to the section in which they are embedded. The contribution of the floating wire to the series capacitance is not as much as interleaved winding, but its labor cost is less. A third method is to interleave sections to achieve a nearly linear distribution under impulse voltage throughout the whole winding. The disadvantage of this method is that the labor cost of interleaving sections is high. For a large rating and high-voltage unit, it is worth to interleave a few sections to achieve satisfied results.

It should be mentioned that besides sections near line end, the sections near the neutral end of some windings may also have severe voltage gradient, as shown in Figure 4.3c which is the maximum voltage of each section of a real unit under full and chopped waves. Measures such as the insertion of floating wire may be applied to mitigate this steep drop.

Also, in the case of some tap sections not being in circuit when the unit is on other than maximum tap, the overhang sections may experience a high voltage at their loose ends.

Although the winding resistance is not considered in Equation (4.3), both its existence and the core loss damps the oscillation. The oscillation wave is also damped by the network; such external damp has a factor about 1.25 [3].

The purpose of the analysis just conducted on an ideal model is to express the common features of voltage oscillation under an impulse by a simple method. The differences between the ideal model and practical winding are, first, real transformer winding consists of a finite number of elements, not an infinite element. Second, real transformer windings are not homogeneous; for example, sections next to line terminals often have fewer turns than sections in the middle of winding, and the spacer thickness may vary along the winding height. Such inhomogeneous affects the voltage distribution in such an inhomogeneous area. Third, a real transformer has winding resistance and core loss, which damp the voltage oscillation. However, at the beginning of the voltage oscillations, the effect of damping is hardly noticeable; therefore, it generally has no influence on the maximum voltage which determines the winding insulation. Fourth, a real transformer has several windings on one leg, and mutual inductances exist between the windings. Such mutual effect is not considered in the analysis on the ideal transformer. In some cases, these mutual inductances cause severe voltage oscillation. Fifth, since a real transformer has more than one winding, when impulsing one winding, there is a transferred overvoltage in other windings; the analysis method just described is inapplicable in such a case. Finally, unlike step voltage, the impulse voltages on real transformers have finite front time and tail.

4.2.2 TRANSFERRED VOLTAGE

For two or more windings on one core leg, when one winding is under impulse, due to capacitive and inductive coupling, transferred voltages are generated in other windings. In some cases, the transferred voltage is so high that it causes insulation failure. The transferred voltage contains two components, inductive and capacitive. Capacitive transferred voltage, V_c, can be estimated as

$$V_c = K_c \cdot \frac{C_2}{C_1 + C_2} \cdot V \tag{4.11}$$

where V_c is the voltage transferred to concerned winding due to capacitive coupling. C_1 is the capacitance per unit length between core and the winding concerned, and C_2 is the capacitance between the winding concerned and the winding impulsed with amplitude V. K_c is a margin factor considering possible oscillation; it is suggested to be 1.25 for general cases, 2.0 when HV is interleaved winding and 2.5 when LV winding has opposite direction to HV winding. If the number of turns in LV winding is approximate to the number of turns in HV winding, factor 2.0 may be chosen. The capacitive transferred voltage becomes a serious problem when HV winding line end is adjacent to a floating end of LV winding, such as a single-phase transformer having 2-leg core, LV windings on each leg are connected in series. Another case is the neutral of wye-connected LV winding being connected to the series winding of a series transformer.

The capacitive transferred voltage is affected by grounding and grounding location of the un-impulsed winding. Five cases in which the un-impulsed winding has different grounding location are shown in Figure 4.4, and their transferred voltages are studied [3]. The transferred voltage to the ground (v_{max}) is highest in cases (a) and (b), while the transferred voltage imposed on one winding element/section (ΔV_{max}) reaches maximum in cases (c) and (e). When unit has delta-connected stabilizing (buried tertiary) winding, and this delta has only one corner grounded, which means one winding has both its ends floating like case (a), two windings have one end grounded like case (b), The capacitive transferred voltage is about 0.4 pu. Some manufacturers set 0.7 pu as standard to keep a safety margin.

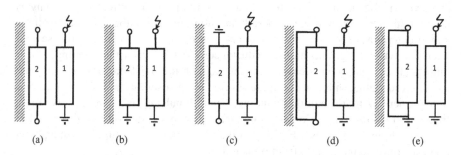

FIGURE 4.4 Transferred voltage initial distributions.

Another transferred voltage comes from inductive coupling; due to the existence of the core, the un-impulsed windings could get voltage when other winding is impulsed. This inductive transferred voltage, V_l, can be estimated as

$$V_l = K_l \cdot \frac{N_L}{N_H} \cdot V \tag{4.12}$$

where N_L is number of turns of winding un-impulsed, N_H is the number of turns of winding impulsed, K_l is a factor considering possible oscillation. It is seen from this equation that the higher the number of turns of un-impulsed winding, the higher its transferred voltage.

The voltage oscillations can also develop along un-impulsed winding, around its steady-state voltage distribution defined by turn ratio. The maximum voltage from oscillation, V_{max}, is

$$|V_{max}| = |V_l| + |V_l - V_c| \tag{4.13}$$

Keep in mind that the inductive transferred voltage strongly depends on the polarities of both windings. If two windings' polarity are the same, $|V_l - V_c| = V_l - V_c$. If the polarities are opposed to each other, $|V_l - V_c| = V_l + V_c$.

In the case when voltage transferring takes place from HV winding to LV winding, since V_l is relatively low, by adjusting the capacitance properly it is possible to avoid the oscillation ($V_l = V_c$) in the LV winding, so the transferred voltage can be controlled low. In the case when voltage transferring takes place from LV winding to HV winding, V_l is relatively higher, V_c is generally much lower than V_l, the oscillation in HV winding cannot be eliminated by modifying the capacitance, and the transferred voltage is high. In some autotransformers, a dangerous overvoltage can appear at HV terminals under not only lightning impulse, but also switching surge.

Transferred voltage occurs not only between windings on the same leg, but also between windings on different legs. A typical case is the three-phase three-leg core unit. HV winding is wye-connected, during a switching impulse test with neutral grounded, the un-impulsed open terminals get 0.5 pu voltage by turn ratio, which makes voltage between the impulsed terminal and the un-impulsed terminal 1.5 pu. The capacitive transferred voltage superimposed on the inductive voltage is sometimes as high as 0.3 pu, which makes the total phase-to-phase voltage about 1.8 pu.

When HV line lead is brought out from the center of the winding, the voltage between HV and LV windings at the center location may be as high as 1.2 pu of impulse voltage due to the transferred voltage built in LV winding. This overvoltage should be considered in the design stage.

4.2.3 Voltage across Regulating Winding

When HV winding or LV winding has a separate regulating winding for voltage regulating, voltage of regulating winding to ground and voltage between main and regulating windings need attention. Commonly used connections of regulating winding to main winding are shown in Figure 4.5. The connection could be linear, plus/

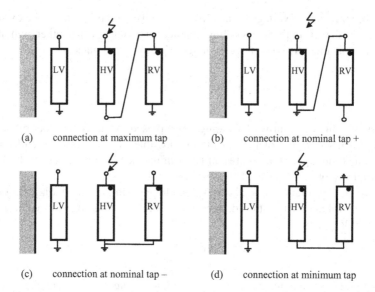

(a)   connection at maximum tap   (b)   connection at nominal tap +

(c)   connection at nominal tap –   (d)   connection at minimum tap

FIGURE 4.5 HV tap winding connection.

minus, or coarse/fine. The type of the regulating winding could be multi-start, interleaved disc or continuous disc. The voltage of the main winding is increased by connection (a), decreased by connection (d) and experiences no change in connections (b) and (c).

The voltage across the gap between the HV and regulating windings, U_{gap}, is the highest in case (d), which is illustrated in Figure 4.6; the highest voltage is above level of impulse voltage applied to HV terminal. The highest voltage of the regulating winding sometimes exceeds twice the voltage proportional to the number of turns. The insulation system should be designed to this maximum gap voltage and the maximum voltage on regulating winding. The voltage across the regulating winding can be estimated by Equation (4.12), in this case $K_l = 2\sim3$. The previous discussion is based on HV winding having a separate regulating winding; the same method is applied to LV winding having a separate regulating winding.

Example 4.1

An arrangement of HV coil and its two tap coils is as shown in Figure 4.7. The HV coil has 167 turns, and each tap coil has 45 turns. The HV line is impulsed with 650 kV BIL (Basic Impulse Level), the inductive transferred voltage on each tap coil is 650 × (45/167) = 175 kV. The two tap coils are connected with opposite polarity. As the result, the voltage in the gap between the two tap coils is the voltage across Tap coil 1 plus the voltage across Tap coil 2. Taking the capacitive transferred voltage as half of the inductive transferred voltage, the estimated voltage in the gap is 175 × 2 × 1.5 = 525 kV. The gap thickness should be decided by this voltage.

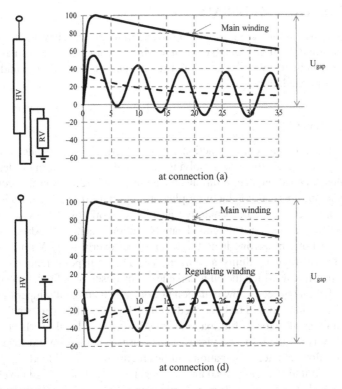

at connection (a)

at connection (d)

FIGURE 4.6 Voltage across main and regulating windings.

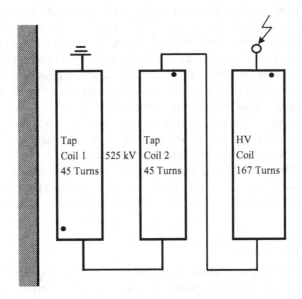

FIGURE 4.7 Arrangement of HV coil with its tap coils.

4.3 INSULATION MATERIALS

Two types of insulation material are used inside a transformer. They are liquid insulation such as mineral oil or vegetable oil, and solid insulation such as insulation paper on winding conductor, winding axial and radial spacers, lead support and winding top and bottom insulation.

4.3.1 MINERAL OIL

Oil in a transformer has several functions, such as thermal cooling, electrical insulating, helping extinguish arcs and dissolving gases and moisture from oil degradation and cellulose insulation deterioration and atmosphere. As a thermal coolant, its specific heat and thermal conductivity play an important role in heat transfer. Heat generated in the core and coil is mainly transferred by oil flow. Low viscosity offers good oil flowing, and as a result, good heat dissipation is achieved. Also, low viscosity assists oil flow in narrow ducts to prevent local overheating. With temperature decrease, the oil-flowing ability also decreases. The pour point temperature is the lowest temperature at which the oil is capable of any observable flow; units operating in very cold climates need a low pour point of oil. When ambient temperature is low, such as −50°C, units should be energized with no load or with a very small load first for a certain period before a full load is applied. This way local overheating due to poor oil flow is avoided, the oil viscosity decreases gradually by no-load loss and improves oil flow and heat dissipation abilities required for a full load. Another low-temperature limitation is at the load tap changer. The operation of load tap changer is usually blocked at −25°C to −40°C depending on tap changer design and type of oil used. Most utilities, with few exceptions, specify that a tap changer be capable of operation at −40°C.

Like other mineral oil, transformer oil has potential risk of fire or explosion. In order to avoid such risk, it is necessary to operate transformer with oil temperature lower than oil flash point. In practical operation, oil temperature is usually much lower than oil flash point, because the process of mineral oil decomposition starts to accelerate considerably at 130°C, further, the thermal decomposition of cellulose and other solid insulation happens at much lower temperature than that for oil decomposition, and at a rate exponentially to the temperature.

Ideal transformer oil has low viscosity, a low pour point, a high flash point, excellent chemical stability to resist to oxidation, and high dielectric strength. However, there is no such oil on earth, so compromise has to be made.

As electrical insulation, transformer oil needs to have high dielectric strength or electrical withstand voltage. The dielectric strength of the insulation structure of a transformer depends not only on the oil dielectric strength but also on the factors discussed next.

4.3.1.1 Electrode Shape

The more uniform field generated by the electrode, the higher the dielectric strength. This is why electrical conductors such as winding conductors have a round edge and application of bushing bottom shield when voltage is high. When electrodes with sharp edges have to be used, such as bolt and nut, they are covered by a round shield when the voltage of parts nearby is high.

4.3.1.2 Electrode Spacing

Increasing the distance between electrodes, for example between windings, gives rise to dielectric strength of the winding system. However, it is not economical because it usually increases amount of copper and the size of the transformer, and an increased space between electrodes doesn't increase the strength proportionally. Empirically, the strength varies closely with the electrode spacing raised to the 2/3 power. However, this relation does not hold true for all electrode shapes, and it definitely doesn't apply to near zero electrode spacing.

4.3.1.3 Electrode Area

The dielectric strength of the oil decreases with increased electrode surface area.

4.3.1.4 Duration of Applied Voltage

The experiments and tests have verified that oil dielectric strength is strongly dependent on the duration of applied voltage and frequency. The oil can take much higher voltage in short duration such as impulse, or lower voltage in a long term such as rated service voltage. A unit has to be designed to take all of types of voltage tests like full wave, chopped wave, switching surge, enhancement and induced voltages, and power frequency applied voltage, each of which has a different shape, voltage level and duration. Selecting only one of these voltages as the design base puts in doubt the credibility of the selection. In order to cope with all types of test voltages, the concept of design insulation level (DIL) factor is introduced: divided by DIL factor, the range of which is listed in Table 4.1, the level of each test voltage is equivalent to the power frequency applied voltage test level, which is called DIL level of this test voltage. For instance, HV winding of a unit will go through 750 kV full wave, 825 kV chopped wave, 620 kV switching surge, 345 kV applied voltage for one minute, 240 kV enhanced level for 7200 cycles and 210 kV one hour induced level. For full wave, its DIL level is 750/2.5 = 300 kV, for chopped wave 825/2.65 = 311.3 kV, for switching surge 620/1.9 = 326.3 kV, for applied voltage 345/1.0 = 345 kV, for enhanced induced 240/1.0 = 240 kV and for one hour induced 210/0.8 = 262.5 kV. The insulation of HV winding to ground should be designed to the highest level of these equivalents, i.e., 345 kV.

TABLE 4.1
DIL Factors

Full wave	Chopped Wave	Switching Surge	Enhanced Level	One hour Level	Applied
2.5	2.5~2.7	1.8~2.1	1.0	0.8~0.9	1.0

4.3.1.5 Temperature

The oil dielectric strength increases with temperature from −5°C to 80~100°C, above which it reduces. Below −5°C, the strength increases rapidly as moisture in suspension in oil get frozen.

4.3.1.6 Oil Volume

There is a special dielectric phenomenon in oil called volume effect. The phenomenon is that the dielectric strength of a smaller volume is considerably higher than the dielectric strength measured on a large volume of oil. The major insulations such as gap between high voltage and low voltage windings, end clearances between windings and yokes take advantage of this by dividing large oil volume into small volume to increase dielectric strength. As a result, the sizes of core and coil are reduced.

4.3.1.7 Velocity

The dielectric strength of oil flowing with a velocity of about 10 cm/s is slightly higher than stationary oil, since flowing oil sweeps away the particles which could otherwise bridge the electrodes making a weak link. With higher flow speed, however, the dielectric strength decreases, because the number of particles passing between electrodes increases. Increasing the flow speed from 10 cm/s to 200 cm/s may cause a drop of 20~25% of the dielectric strength [3]. For forced oil circulation units, the oil flow speed needs to be designed carefully to avoid much drop of oil dielectric strength.

4.3.1.8 Moisture

The dielectric strength is strongly dependent on the state of moisture in oil, i.e., emulsion or solution [6]. When the moisture is in emulsion state, which is of tiny water drops in oil, the lowest breakdown voltages are measured. When the moisture is in solution state in oil, the breakdown voltage is higher, the level of which depends on how much moisture can be dissolved by oil. Oil solubility is temperature dependent; at low temperature, oil solubility is low, emulsion states of moisture extend and the oil breakdown voltage decreases. With increase of oil temperature, the solubility increases, solution states of moisture extend and the breakdown voltage increases. This explains why the oil breakdown voltage is temperature dependent. The temperature of the oil in a transformer tank varies with the tank height; it is lowest at the tank's bottom part, highest in the top part. As a result, the oil in the top part is driest, having the highest breakdown voltage, while the oil at the bottom part is wettest, having the lowest breakdown voltage, where the highest electrical stress is usually located. A study shows that most service-aged units fail in the lower one-third of the windings, which is in the wettest area [7].

Another hazardous impact of high moisture content occurs when the oil contains foreign particles, such as long fiber (e.g. piece of paper or yarn). The fibers align in the direction of the electric field, in the presence of moisture, building a conductive bridge between different potential electrodes, and resulting in a flashover which may severely damage the transformer.

The water can be residual water left after the vapor phase or final vacuum dry processes, or from the atmosphere when the transformer is opened for inspection, or ingress through leakage channels. It is also generated by degradation of solid insulation. Beside two existing forms of moisture in oil as discussed earlier, the water can also exist in a transformer as free water at the bottom of the tank, ice at the bottom if the oil-specific gravity is less than 0.9.

Paper insulation has a greater affinity for water than oil does. The paper insulations hold much more water than oil. The water content in paper is temperature dependent; for example, at 60°C, water in paper is about 300 times of water in oil, at 40°C it is about 1000 times, at 20°C it is about 3000 times. Raising temperature can extract water out of paper insulation, so hot oil circulations are sometimes used to reduce water content in the paper. Note that the water content in an oil test result of a transformer shows only water in the oil, not water in the paper insulation, and it is only a small portion of total content of water in the unit.

In addition to the negative effects of moisture on oil dielectric strength, each time when the moisture is doubled in the unit, the life of the insulation is cut by one-half [7]. The co-existence of both moisture and oxygen is extremely hazardous to transformer insulation, as they accelerate paper insulation decay, form acid and produce sludge and more moisture. The preventive measure is to keep the transformer as dry and free of oxygen as possible.

4.3.1.9 Gas in Oil [3,8]

Depending on its state in oil, gas affects the oil dielectric strength differently. When the gas is dissolved in oil, its effect is the minimum. When the gas is bubbling either in microscopic or visible form, it reduces the oil dielectric strength. The presence of gases accelerates the oil aging process, too. By analyzing gases in oil, three main types of defects may be detected. High hydrogen content may come from partial discharge, high acetylene from sparking and arcing, high ethylene from overheating.

When they are totally dissolved in oil, gases don't cause much oil dielectric strength reduction. Gases become hazardous only when they are in the states of bubbles in oil. Mineral oil can dissolve a limited amount of gases; above this limit, the gases exist in the state of bubble. The solubilities of gases are dependent upon pressure on it and the temperature; the oil dielectric strength therefore is a function of the pressure and temperature. With increase of applied pressure on it, the oil solubilities increase, and greater amounts of gases change the states from bubble to dissolved. This may explain the phenomenon that when a unit has tested a high partial discharge level, increasing and keeping pressure on the oil for a certain period, the partial discharge level reduces, since the number of bubbles in the oil is reduced under the higher pressure. It also indicates that oil is not processed well.

4.3.1.10 Oil Oxidation

All mineral oils are subjected to attack by oxygen and moisture. They make oil oxidation and sludge in oil, increase in acidity of the oil. The oil sludge could cover core and coil, block oil ducts to reduce oil circulation. As a result, core and coil temperature rises further, the oxidation is accelerated and more sludge is made. For each

8~10°C increase in temperature, the oxidation rate will be approximately double. More sludge makes the temperature increase further, which may eventually cause the transformer to fail. Another result of oil oxidation is the increase in acidity of the oil. The acidity causes corrosion and accelerates the degradation of solid insulation. Copper is a strong catalyst in the oil oxidation process, while iron is also a catalyst but not quite so strong. There has been a tendency to increase operating temperature for cost saving, particularly in distribution transformers, the oxidation may be an issue. The methods to minimize oil oxidation is to seal oil from the atmosphere by application of a sealed tank or a tank with a conservator. Another method is adding oxidation inhibitor, such as ditertiary butyl paracresol (DBPC), into the oil; this can reduce oxidation of cellulose insulation and oil. The oxygen attacks the inhibitor instead of the cellulose insulation and oil, and as the transformer ages, the inhibitor is used up and needs to be refilled.

4.3.2 Natural Ester Liquid [9, 10]

Two types of vegetable oils, FR3 of Cooper Power System and BIOTEMO of ABB, are available on the market. They are made of food-grade vegetable oils and food-grade performance enhancing additives. Vegetable oils have certain advantages over the mineral oil as well as certain properties that are inferior. The first advantage of vegetable oil is its biodegradability; it is 97% or greater comparing to 23~35% of mineral oil, which means it is friendly to the environment.

4.3.2.1 Fire Safety

The oils developed by ABB, natural and synthetic esters have a quite high flash and fire points. The synthetic ester exceeds 300°C. FR3 has a fire point of 360°C and a flash point of 330°C; these enable the oils to be classified as less flammable dielectric coolant by both FM (Factory Mutual Global) and UL (Underwriters Laboratories). It allows some insurance relief for installation in or near buildings. OSHA (the Occupational Safety and Health Administration) permits FR3 fluid filled transformer to be installed indoors, typically without sprinklers or vaults, with minimum clearance to a wall of 36 inches. Transformers filled with vegetable oil not only provide the proven performance of liquid-filled design, but also at a lower total life cycle cost than other alternatives with equal rating. Based on reference [10], at greater than 7.5% mineral oil contamination, the fire point of FR3 falls below 300°C, so the limits of residual mineral oil should be 3~5% when the unit is retrofilled with FR3.

4.3.2.2 Service Life

Vegetable oils have a greater affinity to hold water with a solubility of about 1200 ppm, as opposed to 60 ppm or so of mineral oils at room temperature. This ability to absorb a large amount of moisture is an advantage in transformer service life by enhancing the paper insulation life, since the oil can preferentially absorb more water from the paper. Tests show that paper aged in FR3 fluid takes 5~8 times longer to reach the same end-of-life points as paper aged in conventional transformer oil. The thermal decomposition by-products from FR3 fluid are essentially limited to CO_2 and H_2O with trace CO depending on the availability of oxygen and temperature.

4.3.2.3 Thermal Performance

Natural esters possess viscosities that are about 4~5 times higher than mineral oil over the range from room temperature to ~100°C, their higher heat conductivity offsets this difference to some extent, but certain transformers may require larger oil ducts. In addition, the pour point of natural ester is relatively high, at around −15°C; this may limit use of the transformer at low temperature places. Manufacturers have tested transformers without pumps to temperatures as low as −70°C and observed essentially no ill effects on energizing the transformers when the fluid was virtually solid. The neutralization number of the natural ester is slightly higher than mineral oil under almost the same conditions. The coefficient of expansion is similar to the conventional transformer oil. Table 4.2 shows the test results of temperature rise of same transformer with FR3 and mineral oil filled respectively.

Dissipation factor is greater than the value of mineral oil. The dielectric strength and most electrical properties of vegetable oils are comparable or greater than mineral oils. In fact, the dielectric constant of natural ester is in the range of 3.1~3.2, whereas that of mineral oil is near 2.2. This should be a significant benefit since the mismatch with paper and pressboard is decreased.

4.3.3 PAPER INSULATION

Kraft paper is widely used in power transformers due to its high dielectric strength, lower power factor and being free from conducing particles. Its main electric properties are listed in Table 4.3. As wire conductor insulation, thermal upgraded Kraft paper is widely used to achieve 120°C insulation temperature in normal service condition.

TABLE 4.2
Temperature Test Results With FR3 as Well as Mineral Oil

	LV winding gradient	HV winding gradient	average oil rise	Top oil rise
FR3	12.1	10.3	38.3	47.1
Mineral oil	9.1	7.4	34.9	41.2

TABLE 4.3
Kraft Paper Properties

Breakdown voltage, power frequency (kV/cm)	Tensile (kN/m)	Loss factor	Resistivity (MΩ cm)	Relative permittivity
100~150	6~8	$<10~30 \times 10^{-4}$	$100~600 \times 10^6$	2~3.8

TABLE 4.4

Oil Impregnated Paper Insulation Properties

Breakdown voltage at power frequency	Relative permittivity
200~400 kV/cm	4~4.5

Another type of paper, crepe paper tape, is used to cover irregular-shaped parts such as connection joints and static ring, since its extensibility enables it to be shaped to irregular contour. In the cases of high temperature rise, NOMEX crepe tape is used. NOMEX paper works at operational temperature up to 220°C. The disadvantage of crepe paper is its tendency to lose elasticity with time, and after years in service, taping of joints may not be as tight as its original state. Heat, water and oxygen are key factors to accelerate paper aging. The aging results in deterioration of mechanical strengths, such as tensile strength. Regarding its dielectric strength, as long as the paper doesn't break mechanically, it keeps satisfactory electric strength. It is said that 6°C rising results in half of its service life. The harmful effects of moisture can be mitigated by reducing the average water content of the paper below 0.5%.

The dielectric strength of oil-impregnated paper depends mainly on the quality of the paper impregnating process and how successful the elimination of effects (of moisture and gas) is, which reduce the insulation strength. Table 4.4 lists the reference value of breakdown voltage and permittivity of oil impregnated paper.

4.3.3.1 Moisture

Unlike water in oil, water in paper insulation reduces paper dielectric properties such as resistance and resistance to partial discharge [3]. Also, water in paper accelerates paper aging; it is found that the life of paper insulation at 120°C is reduced by a factor of 10 when the moisture content increases from 0.1% to 1% [4]. In order to keep a transformer working well for about 40 years, less than 1% water in paper and around 30~40 ppm in oil at 80°C is a reasonable target.

The conductivity of oil-paper system increases slightly between 0 and 1% moisture content and considerably above 1%, so the measured value of the insulation resistance indicates its moisture content.

Water absorbing abilities of oil and paper change in different ways with temperature. When the temperature rises, more water migrates into oil since its absorbing ability rises and that of paper decreases. When the temperature decreases, the opposite process occurs. When it migrates to oil, water sometimes appears in the form of droplets. This is why a sudden load increase may lead to a breakdown. Some [4] think that most of the water in the oil–paper system is located in the paper.

Oxygen and water have a serious detrimental impact on the life of the system, Oxygen is particularly detrimental, as it has been demonstrated that high oxygen content in the system has been the source of excessive CO generation and aging in operating. Suggested limits are 3000 ppm of oxygen and 1% water in the paper.

4.3.3.2 Breakdown Stresses

Different researchers get different breakdown stress results [1,2]. This is understandable because breakdown stresses are sensitive to the experiment set-up, the preparations of the test sample and other factors. The common results are, first, the breakdown stress under AC voltage is inversely proportional to the test sample thickness, the exponent of which is 0.33. Secondly, the dielectric strength of the oil–paper system is affected by the moisture content. At room temperature, the breakdown voltage is practically the same up to 3% moisture content. At 6% paper moisture content, the breakdown voltage is about 80% of the original value. At 80°C, if only dielectric mechanism is active such as under lightning impulse, with 6% moisture content, the breakdown voltage drops to about 50% of its original. If both thermal and dielectric mechanisms are active, such as under a power frequency test, the breakdown voltage is about 15% of the original.

4.3.4 CLAMPING RING

Two types of solid materials are commonly used as winding clamping rings. One type is laminated wood, a synthetic resin-bonded wood board such as beech veneers which are joined together with thermosetting synthetic resin under pressure and heat. Several grades of laminated wood products are available; the main differences between them are density and orientation of the veneer. Another is laminated pressboard; it is made by laminating sheets of high-density pressboard joined by synthetic adhesive. The partial discharge inception voltage, withstand voltage and creepage strength of the pressboard are higher than that of laminated woods.

The reason why the laminated wood has lower partial discharge inception stress is laid to its manufacturing process [8,11]. Due to veneer patch, the voids between the patches are inevitable, as shown in Figure 4.8. When such voids are completely sealed by the adhesive, oil impregnating these voids is hardly possible. Under an applied voltage, the stress in the void is higher than the stress in the oil since relative permittivity of oil is about 2 times air's relative permittivity. The partial discharge occurs first in the voids. As air in voids has lower inception stress of partial discharge, the partial discharge inception level of laminated wood rings is determined by the air in the void, not the oil.

The laminated pressboard has better dielectric strengths than laminated wood, while laminated wood has better mechanical strength than laminated pressboard. If laminated wood is selected as the ring material, the average electrical stresses in the ring as well as along its surface have to be lower, say, less than 2.0 kV/mm at power frequency voltage to prevent occurrence of partial discharge. In order to achieve these, the end insulation clearances of the winding have to be increased; this means increased material costs for core steel, oil and tank steel, as well as unit height which may have shipping limits. In the case that high voltage lead has to go through the ring, the creepage stress along the surface and stress inside the ring have to be checked to ensure the ring has required dielectric strength; laminated pressboard is often used in such a case. Laminated wood and maple wood are also used as lead support, but keep in mind that the voids or imperfects such as wood knot in these materials could cause partial discharge failures.

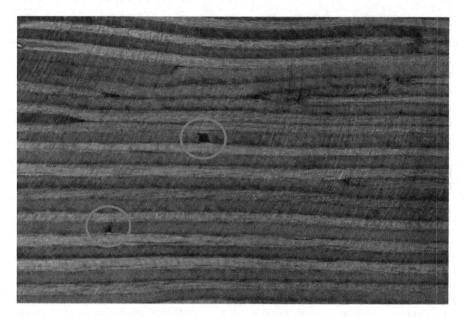

FIGURE 4.8   Side view of a laminated wood.

There are insulation material blocks in difference places in core and coil assembly, such as core blocks beneath the bottom yoke to support the core and on the top yoke between the core and the clamping frame to keep the core in shape, and coil blocks below the bottom ring and at top of the top ring to maintain the pressure exerting on the windings. All of these blocks are in compression, so the flexural strengths of the materials are not a concern; their compressive strengths are so high that they pose no problem. The only concern is electrical stress in and along the surface of the blocks, as the electrical stresses have to be lower than the electrical strengths. For example, Lebanite has low electrical strength; it is an acceptable material for core block since the electrical stress at the bottom of the bottom yoke and the top of the top yoke are quite low. But it is not a good choice for coil blocks because the stress in these areas are relatively high.

4.4   PARTIAL DISCHARGE AND INSULATION STRUCTURE

Good insulation structure is the most important for two reasons. First, it allows the unit to pass all dielectric tests, and the unit may have a long service life. Secondly, it could make the unit overall size smaller with the same voltage class, meaning less material such as core steel, copper and oil to produce the same power [12]. Transformer insulation structure is mainly made by oil ducts of different size and shape. In order to design a robust insulation structure, and to understand dielectric failures as well as efficient utilization of materials and space, it is essential to understand the insulation structures of transformers.

4.4.1 OIL DUCT STRESS

Figure 4.9a is a sketch of two different insulation materials, oil and pressboard, under a uniformed electric field. This field simplifies real electric field between windings in order to achieve simple but useful analysis. The electric stress in each material is

$$E_1 = \frac{V}{\varepsilon_1 d_2 + \varepsilon_2 d_1} \cdot \varepsilon_2; \; E_2 = \frac{V}{\varepsilon_1 d_2 + \varepsilon_2 d_1} \cdot \varepsilon_1 \tag{4.14}$$

where V is applied voltage; ε_1, ε_2 are relative permittivity of oil and pressboard respectively; d_1, d_2 are thickness of oil and pressboard respectively. The relative permittivity of mineral oil is 2.2, of pressboard about 4.4. As the result, the stress in the oil is nearly twice as much as the stress in the pressboard. On the other hand, the power frequency breakdown strength of the oil is about 20 kV/mm while the pressboard is about 40 kV/mm. The oil takes more stress while it has lower breakdown strength; the pressboard takes less stress while it has higher breakdown strength. As can be seen now, the weakness of the whole insulation structure is in the oil, not the pressboard; the breakdown strength of each oil duct determines the whole transformer's insulation structure strength. So, reducing stress or increasing breakdown strength of the oil is a major goal of insulation structure improvement.

Research conducted by Weidmann [13,14] found that the thinner the oil gap, the higher the oil dielectric strength (partial discharge inception stress level), as shown in Figure 4.10. These curves are often called Weidmann curves. In major insulation such as space between high voltage and low voltage windings, and space between winding end and yoke, solid insulation sheets are used to divide bulks of oil into several thin oil gaps, as shown in Figure 4.9b, to increase oil dielectric strength. Here the solid sheets are used not as an insulator but as a divider. It is also found that with a large oil gap, the breakdown voltages scatter greatly; with a thin oil gap, the scattering degree reduces.

By careful selection of the number of oil ducts and duct thickness, to make dielectric strength of each duct is higher than the stress it takes, the probability of the unit passing a dielectric test is higher. Weidmann curves [13] give inception stress of

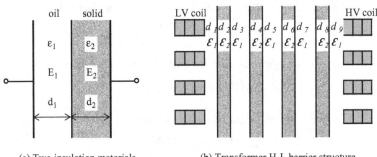

(a) Two insulation materials under uniformed field

(b) Transformer H-L barrier structure

FIGURE 4.9   Basic insulation structures.

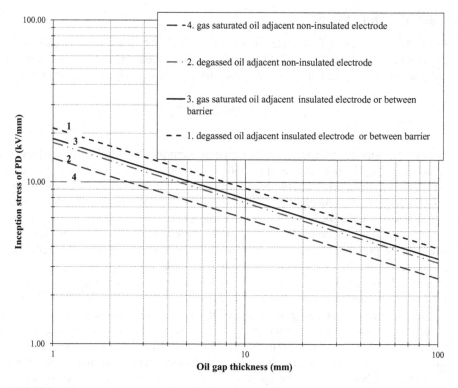

FIGURE 4.10   Inception stress of partial discharge in oil ducts.

partial discharge, the probability of which is about 1% during a one-minute power frequency voltage test, of different oil duct sizes under different oil and electrode conditions. The formula presentations of these curves are

For gas saturated oil, adjacent to non − insulation electrode

$$E_{PD}\left(kV\,/\,mm\right) = 14.0\big/d^{0.37}$$

For degassed oil, adjacent to non − insulation electrode

$$E_{PD}\left(kV\,/\,mm\right) = 17.5\big/d^{0.37}$$ (4.15)

For gas − saturated oil, insulation electrode or between barriers

$$E_{PD}\left(kV\,/\,mm\right) = 18.5\big/d^{0.37}$$

For degessed oil, insulation electrode or between barriers

$$E_{PD}\left(kV\,/\,mm\right) = 21.5\big/d^{0.37}$$

where E_{PD} is inception stress, kV/mm, d is oil duct thickness, mm. From Weidmann curves, the following conclusions may be drawn:

- Subdividing a large oil volume into several thin oil ducts can achieve higher partial discharge inception stress. Keep in mind that excessive solid insulation such as wraps and winding cylinders increases the ratio of solid insulation over oil. The high ratio results in high stress in oil. Considering the wrap's practical mechanical strength and stability, a thickness of 2 mm is a good compromise. Second, thinner oil ducts give higher inception voltages, but it is harder for the oil to flow in the thinner duct due to higher friction, with winding cooling sacrificed as a result. The electrical and thermal performances have to be balanced.

- The partial discharge inception stress in gas-saturated oil is lower by ~20% than one in degassed oil with non-insulated electrode, ~15% with insulated electrode. This indicates that dielectric strength can be improved with a good oil processing. When oil is gas saturated after a unit operates for years, a paper-covered winding conductor helps to keep higher inception stress than a paperless winding conductor. This may be why some utilities want all winding conductors to be paper covered regardless of the winding's voltage.

- The oil ducts between winding and wrap, or between wraps, are maintained by sticks which are placed between them and around the winding or wrap at equal space. The surfaces of these sticks are exposed to tangential electrical field and a possible place for particle deposit, an additional 20% of reduction of the oil strength should be used in design. In practical designs, 20% margin as minimum between stress and strength (partial discharge inception stress) are commonly used.

- Compared to a non-insulated or paperless electrode, the inception stress of an insulated or paper-covered electrode is about 19% higher. It means that covering with paper or placing a barrier adjacent to the bare electrode helps increasing dielectric strength. This is an important measure when the unit's operation voltage is very high, or the space between two different potential electrodes is limited.

In practice, the partial discharge is a major concern since it could deteriorate oil and eventually lead to breakdown. The insulation design should be based on a general rule that the maximum stress in any oil duct is lower than its dielectric strength, i.e. its partial discharge inception stress, by a margin of, say, 20%. Years of accumulated experience have confirmed the validity of Weidmann curves as a designing tool, when they are applied to insulating systems of a moderately non-uniform field distribution, such as encountered in high voltage transformers. The stress limits of Weidmann curves are for 1-minute power frequency applied voltage. Dielectric strength of oil-paper insulation exhibits an exponential decrease when the duration of voltage application increases, and it also changes with wave shape of voltage applied. The transformer has to pass several different types of high voltage tests, such as lightning impulse, switching surge impulse, enhancement induced and one-hour partial discharge, to check its integrity. The duration, wave shape and level of the voltage in each test differ from one another, and the dielectric strength of each such voltage is also different. Based on the experiments [15] and experiences, a group of DIL factors is developed as listed in Table 4.1, to indicate the dielectric strength of each type voltage based on 1-minute power frequency voltage. The DIL factors used by manufacturers are slightly different from each other.

Example 4.2

A 10 mm degassed oil gap adjacent to non-insulated electrode in uniform field, in order to prevent partial discharge from occurring, the maximum stress in the gap under a 1-minute power frequency voltage shall be less than

$$E_{1 minute} = 17.5/10^{0.37} = 7.47 \, (kV/mm)$$

Under full wave impulse and 2.5 DIL factor, the maximum stress shall be below $E_{full wave} = 2.5 \times 7.47 = 18.68$ (kV/mm). Under switch surge impulse and 1.9 DIL factor, the maximum stress shall be below $E_{ss} = 1.9 \times 7.47 = 14.19$ (kV/mm). After years of operation, the oil may be gas saturated, so the stresses are the ones of gas-saturated oil

$$E_{1 minute} = 14.0/10^{0.37} = 5.97 \, (kV/mm)$$

$E_{full wave} = 2.5 \times 5.97 = 14.93$ (kV/mm), $E_{ss} = 1.9 \times 5.97 = 11.34$ (kV/mm). The insulation structure should be designed to withstand all types of voltages during whole life service.

In order to apply Weidmann curves to insulation design in the best way, the following phenomena should be mentioned:

- Weidmann curves are for technical-quality oil. For small oil gaps, the curves prediction is conservative. For a large oil gap, the curves predicted breakdown voltages are also conservative, meaning that measured breakdown voltage (50% breakdown probability) becomes gradually higher than the curve's calculated value with the increasing field non-uniformity. However, for extremely large oil gaps, the reduction of their withstand voltages should be checked due to a very large volume of stressed oil, the volume effect on the breakdown voltage has to be considered. When an extremely large oil volume exists in high electrical stress areas, dividing this volume into several small oil ducts is necessary.
- When sizes of protrusions at electrodes are comparable to the sizes of particles in oil, the electrons streamer generation rate increases with the increasing content of particles. To reduce the protrusion effect on the streamer generation, the electrodes could be covered by insulation material. Further, dividing large volume of oil into several small oil gaps could prevent the streamer propagation. In this way, movements of impurities and particles to high stress area are also blocked.
- The stirring reduces the breakdown voltage significantly, even with small amounts of particles available in the filtered oil. This observation reveals the difference between technical-quality oil circulating in a hot transformer in service, and a sample of the same oil examined in laboratory after a few hours' rest.

It should be noted that Weidmann curves in Figure 4.10 and the calculation method presented earlier are only for a uniform or quasi-uniform field, such as the field between windings; in such a case, the maximum stress is close to average stress. For a non-uniform field such as in winding end's region, the basic principle that dividing big oil volume into thin oil ducts still applies in order to raise the oil dielectric strength, but the method of calculating average stress and comparing it to dielectric strength is dangerous because the local maximum stress can be much higher than average stress. For non-uniform fields, a conservative approach is to limit maximum cumulative stress under dielectric strength given by Weidmann curves. The method to determine cumulative stress in a non-uniform field oil gap is discussed in detail in [14], Figure 4.11a gives an example of cumulative stresses in an oil duct. This cumulative stress is then compared to the dielectric strength of Weidmann curves to ensure that the stresses are always less than the dielectric strength by a good margin. This concept should be applied to determine maximum permissible voltages for non-uniform field gaps.

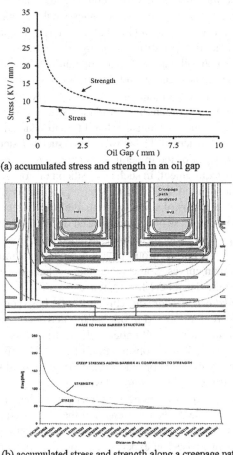

(a) accumulated stress and strength in an oil gap

(b) accumulated stress and strength along a creepage path

FIGURE 4.11   Stress and strength in non-uniform field.

The dielectric strength along a creepage path is analyzed in the same way; an example is shown in Figure 4.11b. First, the tangential stresses along the path are calculated, then a similar method is used to calculate cumulative stresses and compare it to the dielectric strength; here, the dielectric strength should be 70% of values of Weidmann curves shown by the following equation, considering the imperfections at interface.

$$E_{PD,creep}\left(kV \ / \ mm\right) = 16.7 \ / \ d^{0.46} \qquad\qquad (4.16)$$

Weidmann curves provide a reliable tool for calculation of the dielectric strength of the oil gaps in a moderately non-uniform field. A good margin shall be provided to make the unit pass the test and work well in service. A very non-uniform field configuration should be avoided in insulation systems of high voltage power transformers, and an effort to uniform the field distribution as much as possible should be conducted to decrease the local stress. A uniform field also makes efficient use of the oil-filled space inside the tank. This effort could reduce the transformer's overall size, and in many cases the transformer becomes smaller, lighter and less costly owing to reduced volume of copper and insulating material. The main goal of insulation design is to build an oil-paper structure of high dielectric strength, which can be achieved in relatively small oil gaps within a moderately non-uniform field.

It is worth knowing the advantages and drawbacks of an oil-paper insulation system; it is cellulose-based solid insulation materials such as Kraft paper and pressboard with oil. This structure is the predominant insulation structure in power transformers, due to its economic efficiency, technical reliability and adaptation to the technical requirements of the transformer [2]. The drawbacks of such a system are, first, fire risk of mineral oil. The fire point of the mineral oil is about 170°C, so it cannot meet the special non-flammable requirement. In such case, silicon oils or ester liquid have to be used. Secondly, cellulose-based solid materials have permittivity higher than oil. This means the oil has higher electrical stress while its dielectric strength is lower. Beside electrical stress, the aging process of cellulose-based material strongly relates to moisture, oxygen and temperature, which cannot be avoided in service.

At present, there is no one rigorous physical model explaining why inception stresses follow Weidmann curves. Some suggested mechanisms of oil breakdown in transformer insulation are listed here for reference.

Micro-bubble breakdown mechanism. This concept is based on an assumption that the partial discharges are not initiated in oil, but in microscopic gas bubbles that are formed by either decomposition of oil under high electric stress or poor oil processing. This theory is supported by an observation of increased oil dielectric strength with static oil pressure. In practical productions, poor vacuum oil filling often results in partial discharge starting at lower voltage levels.

Surface effect of electrode. It has long been recognized that an insulating coat on metal electrodes increases the dielectric strength of the oil gap adjacent to it; the insulating coat mitigates electrons emissions from the electrode surface protrusions into the oil. For high voltage units, based on experiences, covering metal parts like bolts and nuts by a semi-sphere screen cap and dividing the large oil volume into several small gaps are effective methods to raise the unit's partial discharge inception level.

Effect of particles. The dielectric strength of well-filtered oil is not practically affected by increase of oil volume, but by discharge initiated by protrusions at electrode surfaces. However, with increased particle content, the dielectric strength decreases with the oil volume, and the particles in oil could initiate discharge. The technically clean oil has high enough particle content. The moisture content in oil could also significantly reduce he oil dielectric strength.

Conducting a good production process to remove bubbles, particles and moisture from oil as much as possible, covering the winding conductor with paper and shielding sharp-edged metallic parts are effective methods to improve a unit's dielectric strength.

4.4.2 CORNER STRESS

Besides stresses in oil ducts, corner stress, such as the stress at a corner of a section of a disc winding, has to be below a limit to avoid partial discharge occurring. The experimental data of breakdown stresses of point and rod electrodes reasonably far from a ground plane, which may be taken as reference for corner stress study, are fit to the following equation:

$$E_{positive\ polarity}\left(kV\ /\ mm\right) = \frac{93}{a^{0.35}}; E_{negative\ polarity}\left(kV\ /\ mm\right) = \frac{78}{a^{0.35}} \qquad (4.17)$$

Where a is corner radius, in mm. As can be seen, the breakdown stress at corners where small radii are present can be much higher than the average stress in oil gaps between windings. Table 4.5 lists limits to some of these stresses as a reference.

The corner stress of a winding is also affected by its relative position to adjacent winding. The larger the radial distance between two windings, the less corner stress there is. The effect of axial distance between two windings on the corner stress is shown in Figure 4.12, where A is axial distance from the end of the LV winding to the end of its regulating winding. When the regulating winding is shorter than the LV winding, which is most cases in real designs, A has positive value. When regulating winding is taller than the LV winding, A has negative value. Based on calculated results shown in Figure 4.12, with distance A being positive and increasing, the corner stress of the shorter winding increases, while the corner stress of the taller winding decreases. After passing a certain distance, both corner stresses are stable. With A being a negative value and decreasing, the corner stress of regulating winding decreases while the corner stress of the LV winding increases.

TABLE 4.5
Corner Stress Limits

	Maximum stress allowed at power frequency (kV$_{rms}$/mm)
Winding wire corner insulation surface	11.0~12.0
Bushing shield coating surface with barrier	7.0~11.0
Bushing shield covered by cellulose without barrier	4~4.5
Bare tank wall, core surface	1.5~2.0

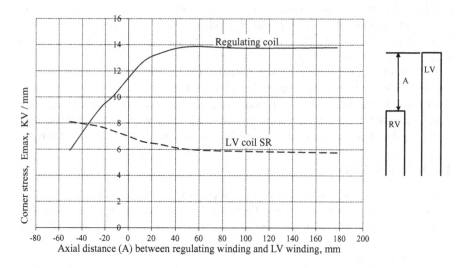

FIGURE 4.12   Relation of corner stress with height difference between windings.

A static ring is sometimes placed at the winding top and bottom to shield the winding corners to reduce these corner stresses. By doing this, the insulation clearance between windings can be reduced. As a result, both size and cost of the unit are reduced. For better use of the static ring, several things need to be kept in mind.

Profile and the insulation thickness of the static ring. Electrical stress on the static ring surface is inversely proportional to its radius; the larger the radius, the lower the surface stress. Also, when the static ring profile fits into one of equipotential lines, the tangential component of the surface stress on the static ring will be reduced to a minimum, thus increasing the static ring creepage strength. However, due to practical limits, it is hard to make such a static ring. The best practical way is to make a static ring the profile of which is close to an equipotential line. Thicker insulation helps on dielectric performances. In the case that a wider static ring is needed to cover big radial build of winding, the static ring is hit by more axial flux and could cause more eddy current and overheating of the static ring. To avoid this, the whole static ring is usually split into several narrow rings.

In an end insulation structures shown in Figure 4.13, the angle ring accompanies the static ring to make an effective insulation. The closer the angle ring is to the static ring, the higher dielectric strength of the oil duct. The angle ring here behaves as a barrier in a non-uniform field; a barrier placed close to the high-stressed electrode can increase the inception voltage of the whole oil gap.

In order to perform its shielding function well, the static ring has to cover over or at least be aligned up with the edges of the winding, such that the winding corners could be shielded. This can be achieved by reducing the number of turns in sections next to the static ring. Further, covering the winding corner with crepe angle could increase the corner strength by about 20%. The static ring on HV winding could increase adjacent winding corner stress if this winding has no static ring.

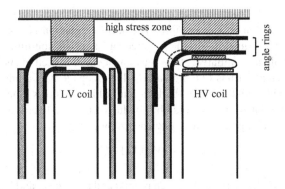

FIGURE 4.13 A top-end insulation.

The corners of the winding top and bottom ends are not the only ones having high stresses; others, such as corners of compensation or tap gaps, may also have high stresses when the gaps are large enough. The same improvement measures as discussed previously are applied to these corners.

The oil gap between the static ring and adjacent winding section decides the capacitance, which affects the impulse performance of the winding. The thinner the oil gap, the higher the capacitance, and the smaller the voltage drop on the section. On the other hand, a thin gap can cause poor oil circulating and high temperature. A 3~4 mm gap is commonly used.

4.4.3 CREEPAGE BREAKDOWN

This type of breakdown happens along the surface of a solid insulator which bridges two electrodes. Some discussion on creepage stress is carried out in Section 4.5.1. Since creepage strength along the surface is lower than insulation puncture breakdown strengths of either oil or solid, it has to be carefully checked during the design stage, especially for high voltage units. Creepage strength has a number of features. First, it goes to its maximum when there is no tangential stress along the surface, and it goes to its minimum when there is only tangential stress along the surface. Second, the creepage strength is material related; pressboard has higher strength than wood or composite wood. This is why pressboard should be used as the lead support for high BIL winding such as 450 kV BIL and above. Third, creepage strength is length related, and shorter length gains higher strength, so dividing a long creepage path into several short paths by application of washer or angle ring can raise creepage strength. Finally, the effect of time duration and frequency of voltage on the creepage strength is not as significant as compared to that on the puncture of solid insulation.

The creepage breakdown stress along the pressboard surface in oil at power frequency voltage can be estimated as follows [16]:

$$E_{cb,ac} = 16.0 - 1.09 \cdot \ln A_c; E_{cb,ac} = 16.6 / d_c^{0.46} \tag{4.18}$$

where $E_{cb,ac}$ is breakdown stress at power frequency, kV/mm; A_c is creepage area, mm²; and d_c is creepage distance, mm. The second equation is very similar to the Weidmann curve in Equation (4.16). The creepage stress is non-uniform along insulation surface path, accumulative stress analysis method is used in such situation. To evaluate this strength, it has to define a possible creepage path along the insulation surface. This path should begin at the location of highest stress along the path. The analysis of accumulative stress should be conducted in both directions. The creepage stress at any point along the path is given as follows:

$$E_{creep} = (V_a - V_b) / d_c \tag{4.19}$$

where E_{creep} is cumulative stress between point a and point b, V_a and V_b are the potentials at points a and b and d_c is the length of path between points a and b. This stress analysis should be conducted along the entire path, not just at the end of the path. Every cumulative stress along any length of the path should be less than the creepage strength by a margin. Note that when the creepage distance is short enough, the mechanism of discharge may change from creepage to streamer.

When the winding height is short enough, there is a possibility of striking along the winding surface. This is a creepage stress issue. To prevent it from happening, a minimum winding height, for example as shown in Table 4.6, should be met.

Considerations for insulation structure design in general are as follows:

- Use insulation barriers to divide large oil volume into small oil ducts to increase oil partial discharge inception voltage.
- Shape molded angle ring to fit into equipotential lines as much as possible to reduce creepage stress.
- Use a static ring in a high stress area to reduce corner stress caused by winding conductor radius, use a screen cap to cover metallic parts which have sharp edges.
- For high voltage winding lead cable, electric stress on its surface sometimes determines its diameter instead of its current carrying capacity.
- Cover the bare conductor with paper if possible.

TABLE 4.6
Minimum Winding Height

BIL (kV)	Height (mm)
350	216
450	279
550	343
650	406
750	470
825	527
900	572
1050	660

4.5 MAJOR INSULATION DESIGN

An insulation design is generally started by deciding the DIL levels of each test voltage the unit will go through. Then, the so-called high-low barrier between windings, end insulation clearances and phase-to-phase clearance are selected based on related DIL levels. It is also decided whether a static ring is needed. Paper insulation thicknesses of winding conductors and shielding pairs are decided be impulse level. Radial spacer thicknesses are determined by impulse level as well as by thermal consideration of oil flow.

4.5.1 MAIN INSULATION GAP BETWEEN WINDINGS IN SAME PHASE

In order to increase partial discharge inception voltage of main insulation gap between windings, pressboard cylinders or wraps are used to divide the main insulation gap into several thin oil ducts, as shown in Figure 4.9b. The partial discharge inception voltage of the thinner oil duct is higher than one of the thicker oil duct, as shown in Figure 4.10. Moreover, the solid insulation such as cylinders and wraps behave as barriers to prevent impurities and particles in oil from establishing long chains between windings, leading to breakdown through the oil. The dielectric strength of the oil duct adjacent to the winding is lower compared to the oil duct between the barriers, due to the presence of the winding conductor's small radius corner and oil gap between sections of the winding. Weidmann curves provide the partial discharge inception voltage of these two types of the duct in degassed oil and gas saturated oil [13].

For simplicity with good approximation, the stresses in the oil duct as well as in solid cylinder or wraps in the middle area between windings can be estimated by uniform field, as follows:

$$E_1 = \frac{V}{\varepsilon_1 \left(\dfrac{d_1}{\varepsilon_1} + \dfrac{d_2}{\varepsilon_2} + \cdots + \dfrac{d_n}{\varepsilon_n} \right)}$$

$$E_2 = E_1 \frac{\varepsilon_1}{\varepsilon_2}, E_3 = E_1 \frac{\varepsilon_1}{\varepsilon_3}, \ldots E_n = E_1 \frac{\varepsilon_1}{\varepsilon_n},$$

(4.20)

where E_i is stress, kV/mm; V is applied voltage level between windings, kV; ε_i is relative permittivity of oil or solid insulation; and d_i is thickness of oil duct or solid insulation, mm. Here are some general design rules for reference.

Designing an insulation gap between two windings is essentially designing the oil ducts and oil-solid insulation interface. For a given gap between two windings, no advantage is gained by increasing the solid insulation thickness in the gap; in fact, more room occupied by solid insulation means less room for oil, which results in higher stress in the oil as discussed previously. The wraps should be as thin as possible, as long as they meet the mechanical strength required in production. For the thickness of the winding cylinder, its mechanical strength is a decisive factor, especially for inner winding. Unit experiencing high forces such as a furnace transformer should have a thicker cylinder. The thickness of oil ducts next to the winding should be decided by both electrical and thermal concerns. If the oil duct is too thin, although its electrical strength is increased, its cooling performance is worsen due to increased friction to oil flowing in a narrow duct. As a result, the winding may get hotter, so a balance has to be achieved. This duct is usually 6 to 10 mm. If the line lead is located at the high voltage winding end, the high-low barrier has to be enlarged to reduce the local stresses in the winding ends. If the line lead is located at the winding center and winding neutral insulation is degraded, the high-low barrier can be reduced. Based on this consideration, the larger high-low gap between the windings is selected for end entry design than that for center entry design [17].

In addition to partial discharge discussed earlier, sticks placed in oil ducts to make duct space raise the chances of creepage failure. The spacing sticks may be stressed by 6~8 kV/mm during the power frequency tests. This case has to be considered during insulation design.

The electric fields at the winding ends are less uniform than the field in the middle area of the high–low barrier between windings, and partial discharge could easily occur in winding end areas. To gain similar dielectric strength as oil ducts in the middle of high-low barrier, the end insulation clearance of winding is usually larger than its high–low barrier clearance. Another reason that the end clearances are larger is that there are high tangential components of stresses along solid insulation in this area, so creepage strengths have to be considered also during design. A good design usually has ~2 kV/mm creepage stress at power frequency along pressboard surface. Dividing the end insulation oil volume into several small oil ducts helps to improve partial discharge inception voltage of the end insulation structure. The detailed discussion is conducted in the following section.

4.5.2 Main Insulation Gap between Innermost Winding and Core

If the test voltage between the innermost winding and the core exceeds a certain level, for example a 185 KV power frequency test, the corners of core steps and tie-plates contribute to the partial discharge significantly, so the partial discharge inception level is related to the gap dimension between the innermost winding and the core, as well as the corners of core steps and tie-plates. When the gap is smaller and the step is thicker, the corner effect is more predominant, and partial discharge from these corners is likely to happen. To prevent such partial discharge, a grounded shield is sometimes used to cover the whole leg. However, making the grounded shield itself partial discharge free and bring its grounding lead out from a high stress area poses a challenge to manufacturers, which could bring failure when they are not made properly.

4.5.3 Main Insulation Gap between Windings in Different Phases

This is called phase-to-phase insulation. The space between phases is decided by the local stresses appearing on the corners of winding's ends or static rings. These stresses are also function of radius and the insulation thickness of the winding conductor or static ring, as well as test voltages. The test voltages under consideration are generally of lightning impulse, induced and switching surge. The switching impulse has some special features listed next.

First, unlike lightning impulses which have such short duration that transformer core doesn't saturate, switching surges happen in a sufficiently long duration that the flux in the core can reach the saturation level. When the flux in the core approaches saturation, the current shows a sudden increase and the voltage rapidly decays to zero. If a higher voltage is applied to winding, the wave tail will be shorter since saturation will set in earlier. No considerable extension of the wave tail is possibly made by increasing energy of the surge generator; the limiting factor is the saturation of the core.

Another difference to lightning impulse is that transferred switching surge voltage appears at open terminals during the test; it consists of a voltage proportional to the turn ratio and a superimposed oscillation. The oscillation may result in considerably high voltage; therefore, the voltage appearing at the non-test open terminals should be checked. For wye-connected windings with the neutral being grounded, without consideration of the oscillation, 1.5 times the test voltage level appears between the terminal of the tested winding and the terminals of non-tested windings for three-phase three-legged core units. In fact, the oscillation always occurs because the existence of capacitance and inductance of windings; around 1.8 times the test voltage level were found in some designs. So the voltages between phases under the switching surge test should be checked, and phase-to-phase insulation should be designed to such a voltage level.

Any minor defect in winding, such as a section-to-section failure, causes a change in the switching surge impedance of the transformer and gives rise to a distortion not only in the wave shape of the neutral current, but also in that of the terminal voltage.

4.5.4 End Insulation

End insulation is the insulation between the winding ends and the yokes of the core. To increase partial discharge inception voltage, the end insulation oil volume is divided into several horizontal smaller ducts, shown in Figure 4.13, as it did in the vertical high–low barrier between windings. Such sub-divided oil gaps are usually larger than the oil gaps in the vertical high–low barrier. There is a zone in which the horizontal oil gaps meet with vertical oil gaps in high–low barriers; the stresses in oil in this zone are usually high and near the winding, and it could cause the oil breakdown. Another issue is the tangential stresses along solid insulation such as angle rings could cause creepage. Application of the static ring can make the electric field in this zone uniform to some extent; as a result the partial discharge inception voltage increases, or the end insulation clearance could be reduced with the same inception voltage level. To reduce the tangential stresses at their surfaces, the profile of angle rings should approximately correspond to the equipotential lines.

The end insulation is also related to clamping structure and winding support structure. The ducts adjacent the winding should not exceed 12.5 mm (~0.5 inch); the rest ducts shall not exceed 24.5 mm (~1.0 inch), except that a duct has to be enlarged for lead's exit. In this case, the related end insulation clearance is also enlarged. From a cooling point of view, the ducts should be large enough to get good oil flow, for which 12.5 mm (~0.5 inch) minimum is preferred. If both LV and HV windings discharge their heat into the same duct, the duct thickness should be 19 mm (~0.75 inch).

As discussed earlier, besides possible striking flashover in oil, creepage is another possible cause of failure due to the existence of the interface and electrical stress along this surface. The field in the end insulation area is non-uniform. The creepage length required under the power frequency test is shown in Figure 4.14. At the beginning of a design, Figure 4.14 may be used as a reference to determine creepage distance; the creepage strength study of the end insulation structure in detail should be conducted when finalizing the design.

When voltages between winding ends and yokes are high, static ring and angle rings are commonly used. More than one angle ring is used when necessary, and the space between two angle rings cannot be too small in order for both angle rings to be effective. In a well-designed end insulation structure, the surface stress along the creepage path under power frequency test voltage, should be kept ~2 kV/mm. Pressboard has better creepage and strike performances than laminated wood. Within limited space under high voltage, useful methods include reducing the oil duct thickness to increase partial discharge inception voltage of oil, increasing the static ring radius to reduce corner stress, placing a thin (~1 mm thick) corner angle ring at winding corners concerned and shortening the barrier length between phases.

Attention is needed for zigzag winding end insulation. During impulsing the neutral terminal, the oscillations between zigzag windings are about 2 times that of applied impulse voltage. The voltage in the gap between zag and HV windings has the same oscillation.

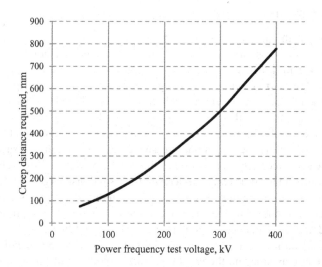

FIGURE 4.14   Creepage distance required in end insulation.

4.6 MINOR INSULATION DESIGN

4.6.1 TURN-TO-TURN INSULATION

Turn-to-turn paper insulation thickness is generally decided by impulse voltage and power frequency test voltage. The winding conductor corner radius and oil quality also play a rule under both voltage tests. When capacitances between turns are small, the voltage distributions under impulse are very non-uniform along the winding. The turns near the end where the impulse voltage enters have very higher voltage gradients between turns than rest of turns in the winding. Interleaving the turns or placing floating shielding conductors in the winding can make such distribution more uniform by adding more series capacitances; typical floating shielding conductor patterns are listed in Table 4.7. Calculation software packages with acceptable accuracy are available which calculate turn-to-turn, section-to-section voltages under impulses. With a good margin to partial discharge inception voltage, a good insulation can be achieved. Keep in mind that the thicker the paper of the cable, the better the turn-to-turn insulation, but the worse the winding temperature gradient and winding space factor.

When floating shielding conductor method is used, turn-to-turn insulation also includes the insulation between the shielding conductor and winding body conductor, the voltage on which is usually higher, for this reason, the shielding conductor needs more paper than the body conductor to get required margin.

Stresses in turn-to-turn insulation are usually high under impulse voltage. Also, the wave shape of the voltages between turns may not be standard wave shape as impulse voltage applied on winding terminal; their duration times are often shorter. To check stresses between turns on common base, the peak voltage levels (V_{peak}) of these waves are corrected to correspondent level ($k \times V_{peak}$) of standard wave where $k = 0.6715 \times t^{0.1017}$ and t is the wave duration, μs.

4.6.2 SECTION-TO-SECTION INSULATION

This duct and winding cable paper form section-to-section insulation. These oil ducts are made by radial spacers placed around the circle of winding sections, as shown in Figure 4.15 which draws only a quarter of the whole winding.

TABLE 4.7
Typical Shielding Conductor Patterns

BIL	Number of turns of shielding conductors in section pairs					
(KV)	1st	2nd	3rd	4th	5th	6th
450	1.5	1	0.5			
550	1.5	1	0.6	0.3		
650	1.6	1	0.6	0.3		
750	1.7	1.2	0.6	0.3		
825	1.8	1.5	1	0.6	0.3	
900	1.8	1.5	1	0.6	0.3	
1050	2	1.8	1.5	1	0.6	0.3

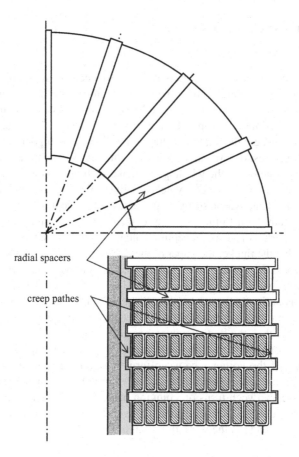

FIGURE 4.15   Radial spacers in winding and possible creepage paths.

The creepage strength along the radial spacer is much lower than the oil striking strength between sections. The creepage strength is the dominant factor of dielectric strength of the section-to-section, so the radial spacer edge radius plays a role.

4.6.3 Tap Gap Location

There are usually two types of gap in winding, tap gap and compensation gap; both are much larger than radial spacer thickness in other places. The tap gap is a gap between two groups of tap sections, the connections of tap sections to the rest of the winding are made through a tap changer. At maximum voltage tap, all the tap sections are connected to the rest of the winding to gain a maximum number of turns; at minimum voltage tap, all the tap sections are off the circuit to make a minimum number of turns. The voltage in tap gap relates the gap location in the winding and number of turns in tap sections. The nearer the line end the gap is, the higher voltage it gets, as shown in Figure 4.16. The more turns the tap sections have, the higher the voltage is. The corner stress of the gap is proportional both to voltage to ground and to the gap thickness under a one-minute applied voltage test; the greater the voltage

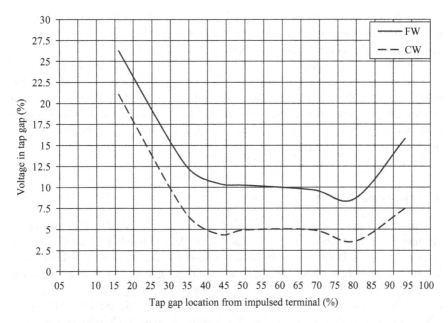

FIGURE 4.16   Relation of tap voltage to its location in winding.

level and the gap thickness are, the higher the corner stress is. However, under the impulse test, the thicker the gap, the lower the corner stress. In order to reduce the stress in tap gap, option 1 is to locate the gap far from the line end. Option 2 is to increase gap thickness to reduce the corner stress under the impulse, and to cover the corner with crepe or molded angle to improve the dielectric strength under both the impulse and one-minute applied voltage tests. Option 3 is to use the angle ring in the gap as a barrier. It is found that shielding conductors in winding reduce not only the stresses between turns and between sections which are adjacent to the impulse termi-nal, but also the stress in the tap gap. Additional pairs of the shielding conductors or longer length of the shielding conductor reduce the stress more, but the stress between the shielding conductor and winding body conductor is increased.

For a unit having a large turn ratio, i.e., the voltage difference between HV and LV windings is significant, it is found that, when impulsing LV winding with HV wind-ing at minimum voltage tap, transferred voltage in tap gap is sometimes so high that breakdown could occur. A similar situation could also happen when voltage tapping ranges are much greater than standard ones.

Compensation gap is the gap between two adjacent sections in a winding and the connection between two sections is permanent. Such a gap is used to balance ampere-turn between windings to mitigate short-circuit forces. The voltage drop in the gap is negligible, however, the corner stress can exceed the inception stress of partial dis-charge when the gap is too big. The gap thickness, high-low barrier between this winding and adjacent winding, and low frequency voltages on winding are key fac-tors on its corner stress. With the same high-low barrier and low frequency voltage, the corner stress gets higher with bigger compensation gap. One way to protect the corner from partial discharge is to use a crepe angle to cover the corner; the inception voltage is improved by about 20%. Another way is to thin out the sections to reduce corner stress; this means placing relatively small gaps in a group of sections instead of placing one big compensation gap between two sections.

A winding with a separated tap winding situated electrically next to the line end may be subjected to an oscillation that requires insulation clearances higher than standard ones. If there is a tap winding next to the neutral end, which is overhanging at some tap positions, there may also be an oscillation and suitable insulation should be provided. In the case of a delta-connected windings, with taps at the line end, during the impulse test two phases are tested simultaneously and there may be a very high oscillation between the phases [18].

4.7 LEAD INSULATION

Winding leads are usually crimped to cable(s) before connected to bushings or tap changer. Lead insulations are ones of lead to lead, lead to windings, lead to tank wall etc. Unfit lead insulation causes failures. Traditionally, the lead is wrapped with insulation tapes to increase dielectric strength, by doing so, the stress of adjacent oil gap is raised, which could result in partial discharge happening in this oil gap. This method could also cause lead overheating. By implementing the principle of dividing a big oil gap into small ones to increase dielectric strength, as shown in Figure 4.17, the lead exit system consists of a group of cylinders centered to the lead, such that both dielectric strength and cooling performance are improved.

Sharp point-to-plane electrode structure is used to analyze non-uniform electrical fields, such as cable lead to tank wall. Without a barrier, the field is extremely non-uniform with very high stress near the sharp point, the partial discharge occurring at sharp point develops towards grounded plane electrode, the electric charged particles generated shorten the clearance, make the dielectric strength of whole system weak. With a barrier, two cases are studied here. In case 1, a barrier is placed in the middle of the oil gap, as shown in Figure 4.18b. The stress near the sharp point is greatly reduced by the sacrifice of increased oil gap stress; however, the overall dielectric strength of the system is improved. This is due to the following: first, as mentioned

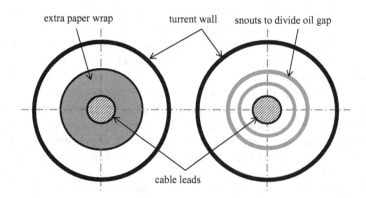

FIGURE 4.17   Lead insulation structure.

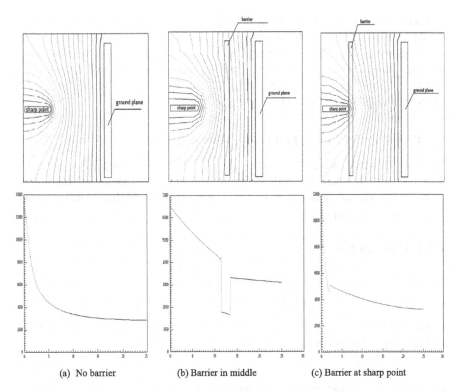

(a) No barrier   (b) Barrier in middle   (c) Barrier at sharp point

FIGURE 4.18 Relation of stress on lead with barrier location.

earlier, the highest local stress is reduced; second, the barrier prevents electric charged particles generated by the sharp point moving to the plane electrode. Finally, the barrier divides the big oil gap into two, which increases the oil gap partial discharge inception voltage. In case 2, as shown in Figure 4.18c, a barrier is placed next to the sharp point. The highest stress occurs in pressboard barrier rather than in the oil; the dielectric strength of the whole structure is also improved since the pressboard has much higher breakdown stress than the oil; in addition, it also prevents the charged particle from movement. In practice, the combination of case 1 and case 2 are often used in high voltage insulation structures.

With the same cable size (conductor diameter and paper thickness), the oil clearances of the cable lead (clr_2) at higher voltage (V_2) can be estimated based on original clearance (clr_1) and voltage (V_1); it is $clr_2 = k_{lead} \cdot clr_1$, where $k_{lead} = (V_2/V_1)^{1.25}$.

When lead cable voltage is high, the cable conductor diameter should be large enough to reduce the stress on it. An experimental formula is used to estimate the minimum conductor diameter, $d = 0.076V + 3.34$, where d is in mm, V is lead system voltage in kV. The estimation is only from voltage consideration; the final section of cable size should consider both the voltage and current carrying ability.

Oil gaps having pieces of solid insulation exist in transformers. For example, in the oil gap between the bushing bottom and the core side frame, there may be lead support structures. The breakdown developed by creepage along solid surfaces should be considered in such cases. To check the dielectric strength of the oil gap, such structure is converted to an equivalent pure oil gap. Then, checking the partial discharge inception voltage, the thickness of equivalent pure oil gap is $L_{oil} + 0.4L_{wood} + 0.6L_{pressbord}$, where L_{oil} is oil thickness, L_{wood} is creepage length of wood and $L_{pressboard}$ is creepage length of pressboard.

4.8 TYPICAL ELECTRIC FIELD PATTERNS

The more uniform a field is, the better use of insulation material that is achieved. However, most practical electrode systems are not uniform. Here are some of the stress calculation equations commonly used; for complex field, finite element analysis method could be applied.

4.8.1 UNIFORM FIELD

$$E_i = \frac{V}{\varepsilon_i \left(\dfrac{d_1}{\varepsilon_1} + \dfrac{d_2}{\varepsilon_2} + \dfrac{d_3}{\varepsilon_3} \right)}; i = 1,2,3 \tag{4.21}$$

where ε_i ($i = 1,2,3$) are permittivity, d_i ($i = 1,2,3$) are dielectric thickness and V is applied voltage, as shown in Figure 4.19.

4.8.2 COAXIAL CYLINDRICAL ELECTRODES

With the inside cylinder being energized by voltage V, outside cylinder grounded, as shown in Figure 4.20, the stress in between is

$$E = \frac{V}{r \cdot \ln\left(\dfrac{R_2}{R_1} \right)} \tag{4.22}$$

The maximum stress is at the inner conductor.

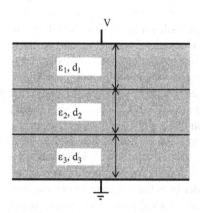

FIGURE 4.19 A uniform field.

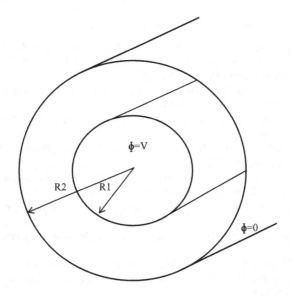

FIGURE 4.20 Coaxial cylindrical field.

4.8.3 CYLINDER TO PLANE

As shown in Figure 4.21, the mean stress is

$$E_{mean} = \frac{V}{(d-R)}$$ (4.23)

The stress at point P and Q are

$$E_p = \frac{V}{d-R} \cdot \frac{\sqrt{(d/R)^2 - 1}}{\ln\left[\sqrt{(d/R)^2 - 1} + (d/R)\right]}$$

$$E_Q = \frac{V}{d-R} \cdot \frac{2 \cdot \sqrt{(d/R-1)(d/R+1)}}{\ln\left[\sqrt{(d/R)^2 - 1} + (d/R)\right]}$$ (4.24)

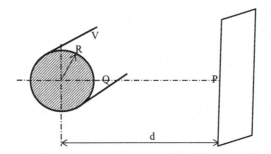

FIGURE 4.21 Rod-to-plane field.

REFERENCES

1. *The Life of a Transformer, 2012, presented by Doble*, Reinhausen, ABB, 2012***.
2. L. F. Blume, et al, *Transformer Engineering,* 2nd edition, John Wiley & Sons, Inc., New York, Chapman & Hall Ltd., London, 1951.
3. K. Karsai, et al, *Large Power Transformers*, Elsevier Science Publishing Company Inc., Amsterdam, Oxford, New York, Tokyo, 1987.
4. Martin J. Heathcote, *The J & P Transformer Book,* 13th edition, Elsevier, Newnes, Amsterdam, et al, 2007.
5. John J. Winders, Jr. *Power Transformers Principles and Applications*, Marcel Dekker, Inc., New York, Basel, 2002.
6. Ravindra Arora, et al. *High Voltage and Electrical Insulation Engineering*, IEEE Press, A John Wiley & Sons, Inc., Hoboken, New Jersey and Canada, 2011.
7. *Maintenance and Troubleshooting of Transformers*, Red dot publication, Columbia, SC, USA, 2017.
8. *Transformer Material Application Guide*, revision 2, Weidmann Electrical Technology, 2001.
9. C. Clair Claiborne, *Natural Ester Based Dielectric Fluids a Manufacture's Development, 15th Annual Doble "Life of a Transformer" Seminar*, February 29-23, 2007, Marlborough, MA, 2007.
10. *Envirotemp FR3™ Fluid*, Bulletin B900-00092 Product information, April, 2005.
11. Bernhard Heinrich, et al., *Issues of Using Laminated Pressboard and Laminated Wood Products in Power Transformers*, 21st International Power System Conference, 46-E-TRN-304, 2006.
12. H.J. Kirch, et al., *Transformer Insulation Engineering*, WICOR Insulation Conference, Rapperswil, Switzerland, September, 1996.
13. H.P. Moser, et al., *Transformerboard,* Special print of *Scientia Electrica*, 1979.
14. V. Dahinden, et al., *Function of Solid Insulation in Transformers*, transform 98, 1998.
15. D.J. Tschudi, *AC Insulation Design*, WICOR Insulation Conference, Rapperswil, Switzerland, September, 1996.
16. Robert M. Del Vecchio, et al., *Transformer Design Principles with Applications to Core-form Power Transformers*, 2nd edition, CRC Press/Taylor & Francis Group, Boca Raton, London, New York, 2010.
17. S.V. Kulkarni, et al., *Transformer Engineering, Design and Practice.*, Marcel Dekker, Inc, New York, Basel, 2004.
18. Jim Fyvie, *Design Aspects of Power Transformers*, Arima Publishing, London, 2009.
19. H.J. Kirch, et al, *Transformer Insulation Engineering, Added Value by Design Optimization, WICOR Insulation Conference*, Rapperswil, Switzerland, September, 1996.

5 Impedances

5.1 POSITIVE SEQUENCE/NEGATIVE SEQUENCE IMPEDANCE

When currents flow through windings of a real transformer, a magnetic field is built up by these winding currents. This magnetic flux field exists outside the core but in the entire space inside tank; it is called the leakage flux field. The action between the leakage flux and windings is expressed by the winding reactance, X. Load loss can be presented by a resistance, R, $Z = R + jX$ is the transformer positive or negative sequence leakage impedance, or simply called impedance. The impedance affects transformer voltage regulation and efficiency as well as short-circuit forces. Higher impedance results in worse regulation, and poor efficiency, while less short-circuit forces, it is a significant character of power transformer. When studying the impedance, exciting current is neglected due to its small value, so that the related mathematical deduction can be simplified with acceptable accuracy. A circuit used to present a two-winding transformer is shown in Figure 5.1. The common way to express the impedance is percentage of voltage drop across the impedance to rated voltage across the whole winding at rated currents. To test the impedance, one set of winding terminals, for example LV bushings, are short-circuited, then a voltage to another set of winding terminals, say HV bushings, is applied to level, V_x, at which its rated current is achieved. The percentage impedance, $Z\%$, is then

$$Z\% = V_x / V_n \times 100 \qquad (5.1)$$

Where V_x is winding voltage at which the rated currents are achieved when another winding is short-circuited, V_n is winding rated voltage. It should be noted that the leakage flux is different from flux in the core. The relation of flux in the core with exciting current is shown in Figure 1.2 of Chapter 1. This relation is nonlinear because of variable permeability of the core steel. The leakage flux, however, is confined in the space outside the core but inside the tank. Most of this space is filled by non-magnetic mediums such as oil, solid insulation materials and winding copper or aluminum which have constant permeabilities, so the leakage flux has a linear relation with load current of windings and does not saturate as the flux in the core does. The return path for the leakage flux is usually either via the route taken through the core for inner winding, or through the tank or tank shield for outer winding. The return path has a much higher relative permeability and therefore has much less influence on the required ampere-turn [1,2].

A certain value of impedance is always required or guaranteed with a tolerance, for example ±7.5% for two-winding unit, ±10% for three-winding unit. Accurate calculation of the impedance to meet guaranteed value is part of design work. The real part of impedance ($Z = R + jX$)—i.e., the resistance—is quite small and easy to calculate. The imaginary part—the reactance—is the focus here. Distribution of ampere-turns along real winding is often non-uniform, such as tapped-out sections in

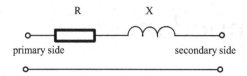

FIGURE 5.1 Simplified equivalent circuit of two winding unit.

HV winding or compensation gap in LV winding where the ampere-turns can be treated as zero. In such cases, advanced analytical techniques or finite element analysis methods which solve Maxwell equations directly are often used to calculate the reactance. In order to indicate influential factors on the reactance, which is important to achieve satisfied value of reactance, a simple calculation method is described here. In this method, uniform distributions of ampere-turns along windings are assumed, meaning there is no gap in windings. It is also assumed that the windings are infinitely long as far as its leakage magnetic field is concerned, to avoid radial flux. A correction factor is added to the final formula for fringing effect of winding ends which is caused by the radial flux.

5.1.1 REACTANCE BETWEEN TWO WINDINGS

The follow deduction is based on the concept that the percentage reactance is the ratio of magnetic energy stored in windings and duct between two windings to transformer rating. Figure 5.2 shows an arrangement of two windings. Since the ampere-turns distribute uniformly along the winding axial direction, ampere circuital law can be simplified as

$$\oint H \cdot dl = \frac{N \cdot I}{h} \tag{5.2}$$

Where H is the magnetic field intensity or magnetizing force, A/m, $(N \cdot I)$ is the ampere-turn of one winding, h is equivalent magnetic winding height, m. The flux density in Tesla, B, is

$$B = \mu_0 \cdot \mu \cdot H \tag{5.3}$$

Where $\mu_0 = 4\pi \times 10^{-7}$ H/m, $\mu = 1$ for non-magnetic medium. The flux density in the gap, B_{12}, is then

$$B_{12} = \frac{\mu_0 \cdot N \cdot I}{h} \tag{5.4}$$

The magnetic energy stored in a non-magnetic medium volume ($\mu = 1$) of V is

$$dW = \frac{1}{2} B \cdot H = \frac{1}{2\mu_0} B^2; \ W = \frac{1}{2\mu_0} \int_V B^2 dV \tag{5.5}$$

Where W is magnetic energy in Watt. The relation of the magnetic energy to inductance, L, is

$$W = \frac{1}{2} L \cdot I^2; \ L = \frac{2W}{I^2} \tag{5.6}$$

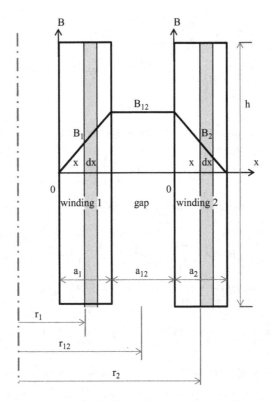

FIGURE 5.2 Leakage flux density distribution between two windings.

The related reactance $X = 2\pi \cdot f \cdot L$, where f is frequency, Hz. The flux density in winding 1, B_1, has a linear relation with the load current, as shown in Figure 5.2, since the mediums of oil, solid insulation and copper or aluminum are non-magnetic:

$$B_1 = \frac{B_{12}}{a_1} x \tag{5.7}$$

Where a_1 is the radial build of winding 1, m. The magnetic energy stored in winding 1, W_1, is

$$W_1 = \frac{1}{2\mu_0} \int_V B_1^2 dV = \frac{1}{2\mu_0} \left(\frac{B_{12}}{a_1} \right)^2 2\pi h \int_0^{a_1} x^2 \left(r_1 - \frac{a_1}{2} + x \right) dx$$

$$= \frac{\pi\mu_0 (NI)^2}{h} a_1 \left(\frac{1}{3} r_1 + \frac{1}{12} a_1 \right) \approx \frac{\pi\mu_0 (NI)^2}{h} \cdot \frac{a_1 \cdot r_1}{3} \tag{5.8}$$

Where r_1 is the average radius of winding 1, m. The magnetic energy stored in the gap between windings 1 and 2, W_{12}, is

$$W_{12} = \frac{1}{2\mu_0} \int_V B_{12}^2 \, dV = \frac{B_{12}^2}{2\mu_0} \cdot V$$

$$= \frac{\mu_0 (NI)^2}{2h^2} \pi \cdot h \left[\left(r_{12} + \frac{a_{12}}{2} \right)^2 - \left(r_{12} - \frac{a_{12}}{2} \right)^2 \right] = \frac{\pi \mu_0 (NI)^2}{h} \cdot r_{12} \cdot a_{12}$$

(5.9)

Where a_{12}, r_{12} are the radial build and average radius of the gap respectively. The flux density in winding 2, B_2, is

$$B_2 = B_{12} \left(1 - \frac{x}{a_2} \right)$$

(5.10)

The magnetic energy stored in winding 2, W_2, is

$$W_2 = \frac{1}{2\mu_0} \int_V B_2^2 \, dV = \frac{B_{12}^2}{2\mu_0} \cdot 2\pi h \int_0^{a_2} \left(1 - \frac{x}{a_2} \right)^2 \left(r_2 - \frac{a_2}{2} + x \right) dx$$

$$= \frac{\pi \mu_0 (NI)^2}{h} \left(\frac{1}{3} r_2 a_2 - \frac{1}{12} a_2^2 \right) \approx \frac{\pi \mu_0 (NI)^2}{h} \cdot \frac{a_2 \cdot r_2}{3}$$

(5.11)

The total magnetic energy stored in the winding assemble, W, is

$$W = W_1 + W_{12} + W_2 = \frac{\pi \mu_0 (NI)^2}{h} \left(\frac{1}{3} r_1 \cdot a_1 + r_{12} \cdot a_{12} + \frac{1}{3} r_2 \cdot a_2 \right)$$

$$= \frac{\pi \mu_0 (NI)^2}{h} \cdot \sum ra$$

(5.12)

Where $\sum ra = \frac{1}{3} r_1 a_1 + r_{12} a_{12} + \frac{1}{3} r_2 a_2$, m^2. Per Equation (5.6), the reactance in ohm, X, is

$$X = \frac{4\pi^2 \mu_0 N^2 f}{h} \sum ra$$

(5.13)

The percentage reactance, $X\%$, is

$$X\% = \frac{V_x}{V} \times 100 = \frac{X \cdot I}{V} \times 100 = \frac{4\pi^2 \mu_0 \cdot N^2 \cdot I \cdot f}{V \cdot h} \sum ra \times 100$$

(5.14)

$$= \frac{4\pi^2 \mu_0}{h \cdot \left(\frac{V}{N} \right)^2} \cdot \frac{\left(\frac{KVA}{\Phi} \right)}{10} \cdot f \cdot \sum ra = 49.61 \times 10^{-7} \cdot \frac{(KVA / \Phi)}{h \cdot \left(\frac{V}{N} \right)^2} \cdot f \cdot \sum ra$$

Where (KVA/Φ) is KVA per phase. V is throughput phase voltage per leg, V/N is the number of turns of the winding. For a two-winding unit, $V;N$ is volts per turn. For engineering convenience, millimeter is often used, and diameter instead of radius is used. As such, Equation (5.14) can be rewritten as

$$X\% = 0.124 \cdot \frac{(KVA/\Phi)}{h \cdot \left(V/N\right)^2} \cdot \frac{f}{50} \cdot \sum Da \tag{5.15}$$

Where h is in mm, and $\sum Da = \frac{1}{3}D_1a_1 + D_{12}a_{12} + \frac{1}{3}D_2a_2$ is in mm². Equation (5.14) or (5.15) has to be modified since in reality, windings have limited length; it means that the leakage flux at the winding ends has a radial component. The Rogowski correction factor, R_g, is used here to modify the reactance for this reason:

$$X\% = 0.124 \frac{\left(\dfrac{KVA}{\Phi}\right)}{h \bullet \left(V/N\right)^2} \cdot \frac{f}{50} \cdot \sum Da \cdot Rg \tag{5.16}$$

Where

$$R_g = 1 - \frac{\sum a}{\pi \cdot h}\left(1 - e^{-\pi \frac{h}{\sum a}}\right)$$

Where $\sum a = a_1 + a_{12} + a_2$. The impedance, $Z\%$, is

$$Z\% = \sqrt{\left(X\%\right)^2 + \left(R\%\right)^2} \tag{5.17}$$

Where $R\%$ is resistance, $R\% = LL/S \times 100$, LL is load loss in Watt and S is rating in VA. $X\%$ can be also expressed as

$$X\% = \frac{reactive\ power\left[KVAr\right]}{rated\ power\left[KVA\right]} \times 100 \tag{5.18}$$

Example 5.1

(a) A two-winding unit has 100 MVA base rating, 60 Hertz, 3 phases. V/N =199.186. LV winding height h_1= 1585 mm, HV winding height h_2= 1575 mm, Mean diameters are D_1 = 1022 mm, D_{12} = 1224mm, D_2 = 1437 mm. Radial builds are, a_1 = 126 mm, a_{12} = 76 mm, a_2 = 137 mm. The average winding height h= (1585 + 1575)/2 = 1580 mm

$$\sum Da = \frac{1}{3} \times 1022 \times 126 + 1224 \times 76 + \frac{1}{3} \times 1437 \times 137 = 201571; \ \sum a = 126 + 76 + 137 = 339$$

Rogowski correction factor and reactance are:

$$R_g = 1 - \frac{339}{\pi \times 1580}\left(1 - e^{-\pi \cdot \frac{1580}{339}}\right) = 0.9317$$

$$X\% = 0.124 \times \frac{\dfrac{100000}{3}}{1580 \times \left(199.186\right)^2} \times \frac{60}{50} * 201571 \times 0.9317 = 14.85\%$$

The calculated load loss P = 204.3 kW, and the calculated resistance is

$$R\% = \frac{204.3 \times 10^3}{100 \times 10^6} \times 100 = 0.204\%$$

The calculated impedance is

$$Z\% = \sqrt{14.85^2 + 0.204^2} = 14.85\%$$

(b) For an auto-connection unit, HV and LV voltages are 230 kV and 115 kV respectively, and (V/N) in Equation (5.15) is not volt per turn as a two-winding conventional connection. V is, for example, HV phase voltage, N is the number of turns of series winding. The impedance between LV and HV windings as two-winding conventional connection with same voltage and MVA rating is $Z_{cvt}\%$ =26.83%. Then Equation (5.15) can be changed as

$$X\% = 0.124 \frac{\left(\dfrac{KVA}{\Phi}\right)}{h \cdot \left(\dfrac{U}{N_s + N_c}\right)^2} \cdot \frac{f}{50} \cdot \sum Da \cdot \left(\frac{N_s}{N_s + N_c}\right)^2 = Z_{cvt}\% \cdot \left(\frac{N_s}{N_s + N_c}\right)^2$$

$$X\% = 26.83\% \times \left(\frac{230 - 115}{230}\right)^2 = 6.71\%$$

Where N_s and N_c are the number of turns in series and common windings respectively.

Under certain rating and frequency, the impedance decreases with increase of $(V/N)^2$ and winding heights, whereas it increases with radial builds of windings and the gap between them. Change of V/N has a more significant impact on the impedance than change of winding shapes. In some designs in which no-load loss and sound level are required to be lower than normal, V/N has to be small in order to reduce core weight; as a result, the impedance is usually high. With a certain V/N, tall and slim windings have lower impedance, while short and fat windings have higher impedance; this is also the direction to change winding shape to adjust the impedance. Increasing the gap or insulation barrier between windings to raise impedance is simple but expansive way; it makes the outer winding larger and increases load loss, and more copper is needed as a result.

5.1.2 Reactance between Series Connected Windings and Other Winding

The complex energy of a transformer is

$$S = P + jQ \tag{5.19}$$

Where P is energy consumed; Q is energy stored which can be expressed as

$$Q = \dot{U} \cdot \overset{*}{I} = -\frac{1}{2}\sum_{k=1}^{n-1}\sum_{j=k+1}^{n} Z_{jk}\left(\dot{I}_j \cdot \overset{*}{I}_k\right) = -\frac{1}{2}\sum_{k=1}^{n-1}\sum_{j=k+1}^{n} Z_{jk}\left(\dot{I}_j \cdot \overset{*}{I}_k + \dot{I}_k \cdot \overset{*}{I}_j\right) \tag{5.20}$$

Where $\dot{I}_j = I_{jr} + j \cdot I_{jj}$; $\overset{*}{I}_j = I_{jr} - j \cdot I_{jj}$; $\dot{I}_k = I_{kr} + j \cdot I_{kj}$; $\overset{*}{I}_k = I_{kr} - j \cdot I_{kj}$ Where I_{jr}, I_{kr} are real parts of the current I_j and I_k, I_{jj}, I_{kj} are imaginary parts of the current I_j and I_k. Equation (5.20) can be re-written as

$$Q = -\sum_{k=1}^{n-1} \sum_{j=k+1}^{n} Z_{kj} \left(I_{kr} \cdot I_{jr} + I_{kj} \cdot I_{jj} \right) \tag{5.21}$$

If Z_{jk} is percentage impedance, and if I_k and I_j are per unit currents, Q is then percentage impedance.

Example 5.2

A unit has 27 MVA base rating, and LV winding has regulating (LTC) winding to adjust its voltage in ±10% range by ±16 steps, as shown in Figure 5.3. Here, reactance is treated as impedance. The impedance of each winding pair at 27 MVA is

$$Z_{34} = 11.75\%; \ Z_{23} = 5.77\%; Z_{24} = 21.81\%$$

At a common base rating, say 27 MVA, the currents in each winding are

$$I_4 = \frac{27 \times 10^3}{\sqrt{3} \times 115} = 135.55\,A$$

$$I_3 = \begin{cases} \dfrac{27 \times 10^3}{\sqrt{3} \times 12.47 \times 1.1} = 1136.43\,A \text{ at maximum tap} \\[2ex] \dfrac{27 \times 10^3}{\sqrt{3} \times 12.47} = 1250.08\,A \text{ at nominal tap} \\[2ex] \dfrac{27 \times 10^3}{\sqrt{3} \times 12.47 \times 0.9} = 1388.97\,A \text{ at minimum tap} \end{cases}$$

$$I_2 = \frac{27 \times 10^3}{\sqrt{3} \times 1.247} = 12500.77\,A$$

Take I_4 per unit current as 1, $I_{4pu} = 1$. At maximum tap,

$I_{4pu} = 1$; $I_{3pu} = -1136.43/1250.08 = -0.909$; $I_{2pu} = -1136.36/12500.77 = -0.0909$

The minus signs of I_{3pu} and I_{2pu} are due to both currents being in opposite direction to I_{4pu}. Base on Equation (5.21), the impedance between HV and LV windings is

$$Z\% = -\left(Z_{23} \cdot I_{2pu} \cdot I_{3pu} + Z_{34} \cdot I_{3pu} \cdot I_{4pu} + Z_{24} \cdot I_{2pu} \cdot I_{4pu} \right)$$

$$Z\% = -5.77 \times 0.0909 \times 0.909 + 11.75 \times 0.909 \times 1.0 + 21.81 \times 0.0909 \times 1.0 = 12.18\%$$

At minimum tap, the per unit currents and impedance between HV and LV windings are

$I_{4pu} = 1$; $I_{3pu} = -1388.97/1250.08 = -1.111$; $I_{2pu} = 1388.97/12500.77 = 0.111$

The plus sign of I_{2pu} is due to the current being in the same direction to I_{4pu}.

$$Z\% = 5.77 \times 1.111 \times 0.111 + 11.75 \times 1.111 \times 1.0 - 21.81 \times 0.111 \times 1.0$$

$$Z\% = 11.34\% \text{ at } 27\,MVA, Z\% = 10.21\% \text{ at } 24.3\,MVA$$

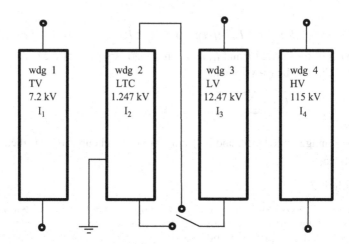

FIGURE 5.3   Winding circuit for impedance calculation.

Another way to calculate impedance between two series-connected windings to a third winding is also presented in Chapter 10. When HV taps are in the main body of the winding, the maximum tap has lowest impedance, while the minimum tap has the highest one. For separate regulating winding, its disposition causes different impedance variations with taps, and so attention should be given when the impedance variations is specified. Table 5.1 lists the tendency of impedance variations, where LV and HV mean LV and HV main windings respectively, LV RV and HV RV mean LV and HV regulating windings respectively.

5.1.3 Reactance of Zigzag Winding

Figure 5.4 shows two zigzag connections of LV winding; the impedance can be calculated through the application of Equation (5.21). Here I_1, I_2, I_3 are the current in HV, Zig and Zag windings respectively, all of them are per unit. Each LV current is 120° apart from another, and the angle between I_2 and I_3 can also be expressed as

$$\dot{I}_3 = e^{j\frac{2\pi}{3}} \cdot \dot{I}_2 = \left(\cos\frac{2\pi}{3} + j\sin\frac{2\pi}{3} \right) \cdot \dot{I}_2 \tag{5.22}$$

Applying Kirchhoff's law,

$$\sum_{k=1}^{n} \dot{I}_k = 0 \tag{5.23}$$

this means that the sum of real currents and the sum of imaginary currents are equal to zero.

$$\sum_{k=1}^{n} I_{kr} = 0; \sum_{k=1}^{n} I_{kj} = 0 \tag{5.24}$$

TABLE 5.1
Relation of Impedance to Dispositions of Windings

Winding Arrangement   **Impedance Variation**

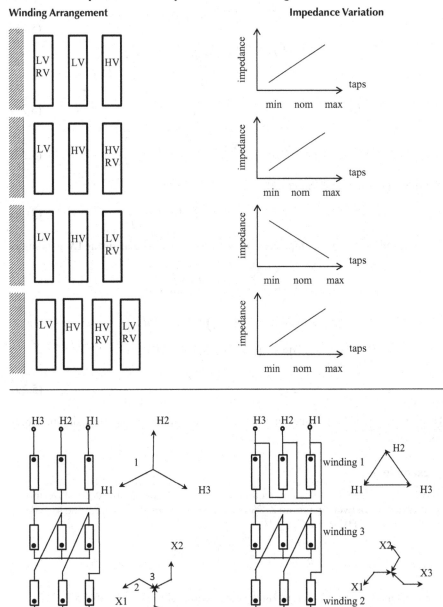

(a) Y - Zigzag, Yz1   (b) D - Zigzag, Dz0

FIGURE 5.4   Zigzag windings.

Where I_{kr} and I_{kj} are real and imaginary currents in each winding respectively. If assuming that $I_1 = 1$, then, $1 + I_{2r} - I_{3r} = 0$; $I_{2j} = I_{3j}$. Based on these and Equation (5.22),

$$I_{3r} + jI_{3j} = \left(\cos\frac{2\pi}{3} + jsin\frac{2\pi}{3} \right)\left(I_{2r} + jI_{2j} \right)$$
$$= \left(-\frac{1}{2}I_{2r} - \frac{\sqrt{3}}{2}I_{2j} \right) + j\left(\frac{\sqrt{3}}{2}I_{2r} - \frac{1}{2}I_{2j} \right) \tag{5.25}$$

$$I_{3r} = -\frac{1}{2}I_{2r} - \frac{\sqrt{3}}{2}I_{2j}; 1 + I_{2r} = -\frac{1}{2}I_{2r} - \frac{\sqrt{3}}{2}I_{2j}; 1 + \frac{3}{2}I_{2r} = -\frac{\sqrt{3}}{2}I_{2j} \tag{5.26}$$

$$I_{3j} = \frac{\sqrt{3}}{2}I_{2r} - \frac{1}{2}I_{2j}; I_{2j} = \frac{\sqrt{3}}{2}I_{2r} - \frac{1}{2}I_{2j}; I_{2j} = \frac{1}{\sqrt{3}}I_{2r} \tag{5.27}$$

From Equation (5.26) and (5.27),

$$I_{2r} = -\frac{1}{2}; I_{2j} = -\frac{1}{2\sqrt{3}}; I_{3r} = \frac{1}{2}; I_{3j} = -\frac{1}{2\sqrt{3}} \tag{5.28}$$

By Equation (5.21), the impedance between HV winding and zigzag LV winding is

$$Z_{1,2+3} = -\sum_{k=1}^{2}\sum_{j=k+1}^{3} Z_{kj}\left(I_{kr} \cdot I_{jr} + I_{kj} \cdot I_{jj} \right)$$

$$Z_{1,2+3} = -\left(Z_{12}\left(I_{1r} \cdot I_{2r} + I_{1j} \cdot I_{2j} \right) + Z_{13}\left(I_{1r} \cdot (-I_{3r}) \right) \right.$$

$$\left. + I_{1j} \cdot (-I_{3j}) \right) + Z_{23}\left(I_{2r} \cdot (-I_{3r}) + I_{2j} \cdot (-I_{3j}) \right)$$

$$Z_{1,2+3} = \frac{Z_{12} + Z_{13}}{2} - \frac{Z_{23}}{6} \tag{5.29}$$

The minus sign of I_{3r}, I_{3j} is because the real current directions are opposite the direction assumed here.

Example 5.3

A 30/40/50 MVA unit has connection as shown in Figure 5.4b. At 30 MVA, the impedances between pairs are $Z_{12}\% = 5.81$; $Z_{13}\% = 3.08$; $Z_{23}\% = 2.22$. The impedance between HV and LV, $Z_{1,2+3}\%$, is

$$Z_{1,2+3}\% = \frac{Z_{12} + Z_{13}}{2} - \frac{Z_{23}}{6} = \frac{5.81 + 3.08}{2} - \frac{2.22}{6} = 4.08\%$$

The reactance of other different zigzag connections are listed in Table 5.2.

5.1.4 REACTANCE OF THREE WINDINGS

The discussion here focuses on transformers which have two LV winding sets which are connected to two LV networks respectively and one HV winding connected to one HV network. Two arrangement types of these two sets of LV windings exist, one

TABLE 5.2
Reactance of Different Zigzag Connections

Connections	Reactance
	For $\theta = 15°$ $$X = 0.7887X_{12} + 0.2113X_{13} - 0.122X_{23}$$ For $\theta = 30°$ $$X = 0.500X_{12} + 0.500X_{13} - 0.167X_{23}$$
	$$X = 1/3(X_{13} + X_{24}) + 1/6(X_{14} + X_{23} - X_{12} - X_{34})$$
	$$X = 1/3(X_{14} + X_{23}) + 1/6(X_{13} + X_{24} - X_{12} - X_{34})$$

Note: Reactance is in percentage.

of which is the radial split, also called low-high-low arrangement, as shown in Figure 5.5a. Another is that one set is stacked on the top of another set, called axial split as shown in Figure 5.5b. The impedances between HV winding and both LV windings, $Z_{H-(X+Y)}$, can be calculated by various methods but should be checked based on Z_{H-X}, Z_{H-Y}, Z_{X-Y} as in the following.

The first step is to get impedance of each branch, Z_H, Z_X, Z_Y, of its equivalent T-network, as shown in Figure 5.5c. These impedances don't exist in reality because impedance always involves at least two windings; the reason to get Z_H, Z_X, Z_Y is to achieve $Z_{H-(X+Y)}$ by mathematical method.

$$Z_H = \frac{Z_{H-X} + Z_{H-Y} - Z_{X-Y}}{2}$$

$$Z_X = \frac{Z_{H-X} + Z_{X-Y} - Z_{H-Y}}{2} \tag{5.30}$$

$$Z_Y = \frac{Z_{H-Y} + Z_{X-Y} - Z_{H-X}}{2}$$

Based on the equivalent impedance network,

$$Z_{H-(X+Y)} = \frac{Z_X \cdot Z_Y}{Z_X + Z_Y} + Z_H \tag{5.31}$$

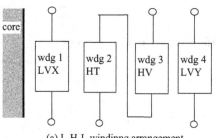

(a) L-H-L windinng arrangement

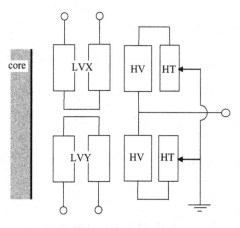

(b) LV axial split arrangement

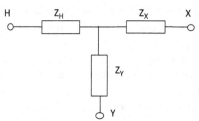

(c) Equivalent impedance network

FIGURE 5.5 Arrangement of two sets of LV winding.

Example 5.4

(a) The calculated impedances of a low-high-low unit as shown in Figure 5.5a at a common MVA are Z_{H-X} = 15.77%, Z_{H-Y} = 12.79%, Z_{X-Y} = 31.19%. Z_H, Z_X and Z_Y are

$$Z_H = (15.77 + 12.79 - 31.19)/2 = -1.32\%; \quad Z_X = (15.77 + 31.19 - 12.79)/$$
$$2 = 17.09\%; \quad Z_Y = (12.79 + 31.19 - 15.77)/2 = 14.11\%$$

$$Z_{H-(X+Y)} = \frac{17.09 \times 14.11}{17.09 + 14.11} - 1.32 = 6.41\%$$

(b) The calculated impedances of an axial split unit as shown in Figure 5.5b are $Z_{H-X} = 7.24\%$ at 21 MVA, $Z_{H-Y} = 8.16\%$ at 15 MVA, $Z_{X-Y} = 12.75\%$ at 15 MVA. At 36 MVA, these impedances are:

$$Z_{H-X} = 7.24 \times (36/21) = 12.41\%$$

$$Z_{H-Y} = 8.16 \times (36/15) = 19.58\%$$

$$Z_{X-Y} = 12.75 \times (36/15) = 30.6\%$$

$$Z_H = (12.41 + 19.58 - 30.6)/2 = 0.70\%; \ Z_X = (12.41 + 30.6 - 19.58)/$$

$$2 = 11.72\%; \ Z_Y = (19.58 + 30.6 - 12.41)/2 = 18.89\%$$

$$Z_{H-(X+Y)} = \frac{11.72 \times 18.89}{11.72 + 18.89} + 0.70 = 7.93\%$$

For axial split winding, as shown in Figure 5.5b, when impedances Z_{H-LVX}, Z_{H-LVY} at half rating, and $Z_{H-(LVX+LVY)}$ at full rating are required to be approximately the same, the axial split winding arrangement is the best fit economically. Assuming the impedance between both LV and HV winding is $Z_{H-(LVX+LVY)}$, the impedance between each LV and HV windings may be estimated as

$$Z_{H-LVX} = Z_{H-LVY} = (0.85 \sim 0.9) \cdot Z_{H-(LVX+LVY)} \tag{5.32}$$

The impedance between LVX and LVY is

$$Z_{X-Y} = 2 \times (0.9 \sim 0.95) \cdot Z_{H-(LVX+LVY)} \tag{5.33}$$

Experiences prove that transformers can be economically designed in a range of reactance, as shown in Table 5.3. For each voltage class, there is a value of impedance which will bring a minimum transformer cost; this impedance is called standard impedance. Variation from 80% to 120% of this value does not raise the cost a lot, as shown in Figure 5.6.

TABLE 5.3

Impedance for Economic Design

BIL (kV)	Voltage Class (kV)	Standard Impedance (%)		
110	15	4.50	–	7
150	25	5.50	–	8
200	34.5	6	–	8
250	46	6.5	–	9
350	67	7	–	10
450	115	8	–	12
550	138	8.5	–	13
650	161	9	–	14
750	230	11.0	–	16

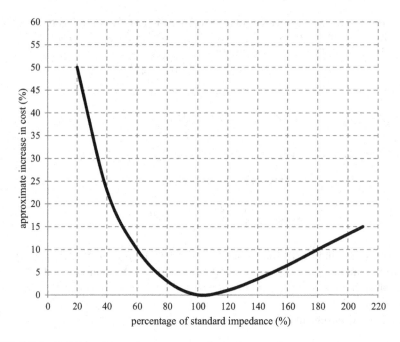

FIGURE 5.6   Relation of impedance to approximate increase in cost.

5.1.5 LEAD REACTANCE

The contribution of high current carrying leads to winding impedance cannot be ignored under some circumstances; for example, low impedance value is guaranteed with tight tolerance, and knowledge of the lead impedance is important to meet this tight tolerance. It is assumed here that the leads are parallel to each other, separated by distance D from center axes of the leads, as shown in Figure 5.7. The lead radius is R. The inductance of this system consists of two parts, self-inductance Li and mutual inductance Lo [3]. The flux density inside the individual lead which relates to Li is

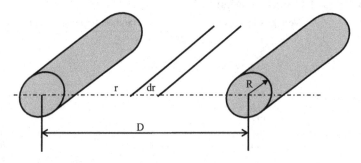

FIGURE 5.7   Lead impedance calculation.

$$B_i = \frac{\mu_0 I}{2\pi r} \cdot \frac{r^2}{R^2} = \frac{\mu_0 I}{2\pi \cdot R^2} \cdot r \tag{5.34}$$

Where r is distance from the center of the cable on left to a point the flux density of which is B_i. The flux through area $h \cdot dr$ is $d\Phi$

$$d\Phi = B_i \cdot h \cdot dr = \frac{\mu_0 Ih}{2\pi R^2} \cdot r \cdot dr \tag{5.35}$$

Where $\mu_0 = 4\pi \times 10^{-7}$ H/m, h is lead length, m. The current interacting with $d\Phi$ is I', not the whole current I. The relation between them is

$$I' = \left(\frac{\pi r^2}{\pi R^2} \right) I = \left(\frac{r^2}{R^2} \right) I \tag{5.36}$$

So the flux which relates L_i is

$$d\Psi = \frac{I'}{I} \cdot d\Phi = \frac{r^2}{R^2} \cdot \frac{\mu_0 \cdot I \cdot h}{2\pi R^2} \cdot r \cdot dr = \frac{\mu_0 \cdot I \cdot h}{2\pi R^4} \cdot r^3 dr$$

$$\Psi = \int_0^R \frac{\mu_0 \cdot I \cdot h}{2\pi R^4} \cdot r^3 dr = \frac{\mu_0 \cdot I \cdot h}{8\pi} \tag{5.37}$$

L_i is then

$$L_i = \frac{\Psi}{I} = \frac{\mu_0 \cdot h}{8\pi} \tag{5.38}$$

L_0 is made by the current in another lead. Since dimension D is usually much greater than the lead radius R, R is ignored when calculating L_0; this means that the currents are concentrated to the lead axes. The flux density between leads is

$$B_o = \frac{\mu_0 \cdot I}{2\pi r} + \frac{\mu_0 \cdot I}{2\pi (D-r)} \tag{5.39}$$

The flux through area $h \cdot dr$ is $d\Phi$:

$$d\Phi = B_o \cdot h \cdot dx \tag{5.40}$$

The total flux through the whole area between the two leads and L_0 are

$$d\Psi = \int_R^{D-R} B_o \cdot dl = \frac{\mu_0 \cdot I \cdot h}{\pi} \cdot \ln \frac{D-R}{R} \tag{5.41}$$

$$Lo = \frac{\Psi}{I} = \frac{\mu_0 \cdot h}{\pi} \cdot \ln \frac{D-R}{R} \approx \frac{\mu_0 \cdot h}{\pi} \cdot \ln \frac{D}{R}$$

The total inductance of the lead system is

$$L = L_i + L_0 = \frac{\mu_0 \cdot h}{8\pi} \times 2 + \frac{\mu_0 \cdot h}{\pi} \cdot \ln \frac{D}{R} = \frac{\mu_0 \cdot h}{\pi} \left(\frac{1}{4} + \ln \frac{D}{R} \right) \tag{5.42}$$

The voltage drop on the reactance X_L is

$$V_X = 2\pi fLI \qquad (5.43)$$

Where f is frequency. The percentage impedance is

$$Z\% = \frac{V_x}{V} \times 100 = 1.51 \times 10^{-2} \times \left(\frac{I}{V}\right) \cdot \left(\frac{f}{60}\right) \cdot h \cdot \left(\frac{1}{4} + \ln\frac{D}{R}\right) \qquad (5.44)$$

Where I is current in lead, A, V is phase voltage, V, and h is the lead length, m.

Example 5.5

A unit has power rating 15 MVA base, 60 Hz. The LV winding is wye connected, line-to-line voltage is 4.16 kV, current is 2082A. The tank height is 4166 mm (164 inch), the core center is 1022 mm (40.25 inch). At the design stage as an approximation, the LV lead length is the same as tank height, and the leads are separated by the same distance as the core center. Each lead consists of four of 800 MCM cable, the radius of which is 16.2 mm (0.6375 inch). The equivalent radius is 16.2 × 2=32.4 mm (0.6375 × 2 = 1.275 inch). Based on Equation (5.44), the lead impedance is

$$Z\% = 1.51 \times 10^{-2} \times \left(\frac{2082}{4160/\sqrt{3}}\right) \times \left(\frac{60}{60}\right) \times 4.166 \times \left(\frac{1}{4} + \ln\frac{40.25}{1.275}\right) = 0.20\%$$

The calculated impedance between HV and LV windings is 8.17%. The total impedance is Z% = 8.17% + 0.20% = 8.37%.

When the numerical value of winding current is more than half of winding voltage, lead impedance contribution to unit impedance is noticeable. In designs in which impedances are low, or the tolerance of impedance is tight, lead impedance should be counted in.

5.2 ZERO SEQUENCE IMPEDANCE [4, 5]

Zero sequence impedance of a three-phase transformer is the impedance per phase between line terminal and its neutral terminal when a zero sequence voltage is applied between joined line terminals and the neutral terminal. This definition can be expressed in the following and is shown in Figure 5.8a:

$$Z_0 = V_0 / I_0 \qquad (5.45)$$

Where V_0 is the voltage between line terminals which are connected together and the neutral terminal and I_0 is the line current. In a three-phase balanced loading system, zero sequence impedances are inactive; only this symmetry condition is interrupted, resulting in an asymmetrical loading condition, the zero sequence impedance is involved.

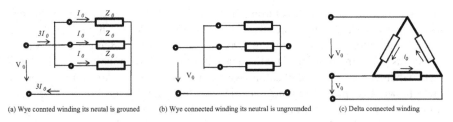

(a) Wye connted winding its neutal is grouned   (b) Wye connected winding its neutral is ungrounded   (c) Delta connected winding

FIGURE 5.8 Zero sequence impedance circuits.

Example 5.6

A transformer has base rating 43.75 MVA unit, HV line-to-line voltage is 230 kV, delta connection. LV line-to-line voltage is 26 kV, wye connection. LV line-to-neutral voltage is $26/\sqrt{3} = 15.011 kV$. LV phase current is $43.75 \times 10^3 / (\sqrt{3} \times 26) = 971.5\,A$. The tested current is $3\,I_0 = 38.6$. The tested voltage is $U_0 = 17.98$ V. The tested zero sequence impedance in percentage is

$$Z_0\% = \frac{17.98/(38.6/3)}{15011/971.5} \times 100 = 9.04\%$$

Zero sequence impedance is line terminal related. It is also winding connection related. Zero sequence current needs a closed-circuit loop to flow. If windings are wye connected but their neutral is not grounded, as shown in Figure 5.8b, these windings don't offer a closed-circuit loop for zero sequence current to flow, and as a result its zero sequence impedance is infinite. Delta-connected windings offer a closed loop for zero sequence current flowing inside the delta, but the zero sequence current never appears in the line, as shown in Figure 5.8c; the result is that the zero sequence impedance is also infinite.

Magnetic circuit or type of core also affects zero sequence impedance. For three-phase three-leg core, under zero sequence voltages, the fluxes in each leg are the same and in phase. Therefore, these fluxes have to leave the core when they meet in yoke, in order to find a return path outside the core. Because reluctance in oil is much higher than the reluctance in the core, a considerable magnetomotive force is required; this means high magnetizing current is required. From an electric circuit point of view, this phenomenon presents low zero sequence magnetizing impedance, as shown in Figure 5.9a. This magnetizing impedance is nonlinear with the current or voltage and varies from design to design. For a three-phase five-leg core, easy return paths are

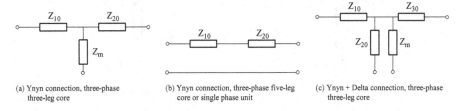

(a) Ynyn connection, three-phase three-leg core   (b) Ynyn connection, three-phase five-leg core or single phase unit   (c) Ynyn + Delta connection, three-phase three-leg core

FIGURE 5.9 Zero sequence impedance networks of Ynyn connection.

available for zero sequence flux through unwounded side legs, so very low magnetizing current is needed. Therefore, the magnetizing impedance is high enough to be treated as infinite. As shown in Figure 5.9b, the open-circuit zero sequence impedance is same as the open-circuit positive sequence impedance. When there is an additional delta winding, the voltage applied on wye winding makes circulating current flow in the delta winding, part of currents in exiting winding balances the circulating current flowing in the delta and the zero sequence impedance network is as shown in Figure 5.9c. In the calculation, the magnetizing impedance, Z_m, is usually neglected.

Also, for three-phase three-leg core, when the zero sequence flux goes outside of the core, this flux interacts with the tank. The contribution of the tank to the impedance in such case is that the tank acts like an outmost phantom delta winding.

Like the positive sequence impedance network, the zero sequence impedances of a transformer can also be presented by a network. The ways to decide the values of each element in the networks are listed in the following. The test methods are also introduced.

5.2.1 TRANSFORMER WITHOUT NEUTRAL TERMINAL

When neither winding has neutral brought out, such as delta windings or wye windings with ungrounded neutral, the impedance to zero sequence line current is infinity, as shown in Figure 5.10a, since the zero sequence line current cannot exist because there is no return circuit loop for it.

5.2.2 TRANSFORMER WITH SINGLE NEUTRAL TERMINAL

The zero sequence impedance can be expressed by a single value, as shown in Figure 5.10b, where the terminal 1 is the terminal of winding which has neutral brought out.

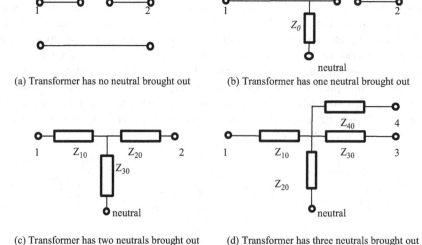

(a) Transformer has no neutral brought out

(b) Transformer has one neutral brought out

(c) Transformer has two neutrals brought out

(d) Transformer has three neutrals brought out

FIGURE 5.10 Zero sequence impedance equivalent circuits.

Test results show that when HV (outer) windings are delta connected and LV (inner) windings are wye connected, the ratio of zero sequence impedance to positive sequence impedance is almost equal to 1. When HV (outer) windings are wye connected and LV (inner) windings are delta connected, the ratio of zero sequence impedance to positive sequence impedance is ~0.87. The tank effect is noticeable.

5.2.3 Transformer with Two Neutral Terminals

The equivalent zero sequence impedance network is as shown in Figure 5.10(c). Three tests, as listed in Table 5.4, are needed to determine the values of Z_{10}, Z_{20} and Z_{30}. From Table 5.4, the following equations are established:

$$\frac{Z_{20} \cdot Z_{30}}{Z_{20} + Z_{30}} = Z_{1NS} - Z_{10} = Z_{1NS} - Z_{1N0} + Z_{30}$$

$$\frac{Z_{20} \cdot Z_{30}}{Z_{20} + Z_{30}} - Z_{30} = Z_{1NS} - Z_{1N0}$$

$$\frac{Z_{20} \cdot Z_{30} - Z_{20} \cdot Z_{30} - Z_{30}^{2}}{Z_{20} + Z_{30}} = Z_{1NS} - Z_{1N0}$$

Then

$$Z_{30} = \sqrt{Z_{2N0}\left(Z_{1N0} - Z_{1NS}\right)}; \; Z_{20} = Z_{2N0} - Z_{30}; \; Z_{10} = Z_{1N0} - Z_{30} \qquad (5.46)$$

TABLE 5.4
Zero sequence Impedance Tests for Transformer Having Two Brought Out Neutrals

	Terminal 1	Terminal 2		
	(HV)	(LV)	(TV)	Test values
Test 1	Short-circuit line terminals apply single-phase voltage between the terminals and its neutral	open	open	$Z_{1N0} = Z_{10} + Z_{30}$
Test 2	Short-circuit line terminals apply single-phase voltage between the terminals and its neutral	Short-circuit line terminals and its neutral	open	$Z_{1NS} = Z_{10} + Z_{20} \, / \! / \, Z_{30}$
Test 3	open	Short-circuit line terminals, apply single-phase voltage between the terminals and its neutral	open	$Z_{2N0} = Z_{20} + Z_{30}$

For a Y–Y connection or an auto-transformer without delta tertiary winding, the zero sequence impedance is ~90% of positive sequence impedance when seeing from HV (outer winding) bushing terminal, ~100% of the positive sequence impedance when seeing from LV (inner winding) bushing terminal. With delta tertiary winding, for Y–Y connection units, $Z_{30} \approx 50\%$ when HV winding neutral is grounded, $Z_{30} \approx 60\%$ when LV neutral is grounded. For an auto-transformer without tertiary winding, Z_{30} is higher.

5.2.4 TRANSFORMER WITH THREE NEUTRAL TERMINALS

The equivalent zero sequence impedance network is as shown in Figure 5.10d. Five tests, as listed in Table 5.5, are needed to get values of Z_{10}, Z_{20}, Z_{30} and Z_{40}

TABLE 5.5

Zero Sequence Impedance Tests for a Transformer Having Three Brought Out Neutrals

	Terminal 1 (HV)	Terminal 2 (LV1)	Terminal 3 (LV2)	Test values
Test 1	Short-circuit line terminals apply single-phase voltage between the terminals and its neutral	open	open	$Z_{1N0} = Z_{10} + Z_{20}$
Test 2	Short-circuit line terminals apply single-phase voltage between the terminals and its neutral	Short-circuit line terminals with its neutral	open	$Z_{1NSX} = Z_{10} + \dfrac{Z_{20} \cdot Z_{30}}{Z_{20} + Z_{30}}$
Test 3	Short-circuit line terminals apply single-phase voltage between the terminals and its neutral	open	Short-circuit line terminals with its neutral	$Z_{1NSY} = Z_{10} + \dfrac{Z_{20} \cdot Z_{40}}{Z_{20} + Z_{40}}$
Test 4	open	Short-circuit line terminals apply single-phase voltage between the terminals and its neutral	open	$Z_{3N0} = Z_{20} + Z_{30}$
Test 5	open	open	Short-circuit line terminals, apply single-phase voltage between the terminals and its neutral	$Z_{4N0} = Z_{20} + Z_{40}$

$$Z_{20} = \sqrt{Z_{20}^2} = \sqrt{(Z_{20} + Z_{30})\frac{Z_{20}^2}{Z_{20} + Z_{30}}} = \sqrt{(Z_{20} + Z_{30})\left[(Z_{10} + Z_{20}) - \left(Z_{10} + \frac{Z_{20} \cdot Z_{30}}{Z_{20} + Z_{30}}\right)\right]}$$

$$Z_{20} = \sqrt{Z_{3N0}(Z_{1N0} - Z_{1NSX})}; Z_{10} = Z_{1N0} - Z_{20}; Z_{30} = Z_{3N0} - Z_{20}; Z_{40} = Z_{4N0} - Z_{20} \quad (5.47)$$

Under symmetrical loading conditions, only positive sequence impedances are active. In the case of asymmetrical loading, or single-phase fault, the system response is decided by both positive sequence impedances and the zero sequence impedances.

Example 5.7

(a) A three-leg three-phase auto unit with delta tertiary winding has positive sequence impedances, $Z_{12+}\% = 6.75$, $Z_{13+}\% = 27.38$, $Z_{23+}\% = 18.5$. From these, the impedances of each branch Z_{1+}, Z_{2+} and Z_{3+}, as shown in Figure 5.11, are calculated as follows:

$$Z_{1+} = \frac{Z_{12+} + Z_{13+} - Z_{23+}}{2} = \frac{6.75 + 27.38 - 18.51}{2} = 7.81\%$$

$$Z_{2+} = \frac{Z_{12+} + Z_{23+} - Z_{13+}}{2} = \frac{6.75 + 18.51 - 27.38}{2} = -1.06\%$$

$$Z_{3+} = (Z_{13+} + Z_{23+} - Z_{12+})/2 = (27.38 + 18.51 - 6.75)/2 = 19.57\%$$

The impedance between HV and parallel connected LV and TV is

$$Z'_{1+} = Z_{1+} + \frac{Z_{2+} \cdot Z_{3+}}{Z_{2+} + Z_{3+}} = 7.81 + \frac{(-1.06) \times 19.57}{-1.06 + 19.57} = 6.69\%$$

The zero sequence impedance Z_{10} is estimated as

$$Z_{10} \approx 0.84 \times Z'_{1+} = 0.84 \times 6.69 = 5.62\%$$

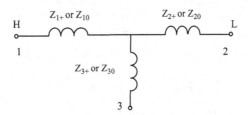

FIGURE 5.11 Impedance network of Y–Y connection unit having a delta winding.

Here, the coefficient of 0.84 is based on tested results of a group of similar units. The zero sequence impedance between HV and LV is estimated as

$$Z_{120} \approx 0.87 \times Z_{12+} = 0.87 \times 6.75 = 5.87\%$$

Here, the coefficient of 0.87 is based on tested results of similar units. Z_{20} is estimated as

$$Z_{20} \approx Z_{120} - Z_{10} = 5.87 - 5.62 = 0.25\%$$

Z_{20} is tested 0.44%. Z_{30} is estimated as

$$Z_{30} \approx 0.81 \times Z_{3+} = 0.81 \times 19.57 = 15.85\%$$

Here, the coefficient of 0.81 is based on the test results of similar units.
   (b) A five-leg three-phase auto unit with delta tertiary winding has tested positive sequence impedances

$Z_{12+} = 4.89\%$, $Z_{13+} = 17.15\%$, $Z_{23+} = 11.20\%$

From these, the impedance of each branch is

$$Z_{1+} = \left(Z_{12+} + Z_{13+} - Z_{23+}\right)/2 = \left(4.89 + 17.15 - 11.2\right)/2 = 5.42\%$$

$$Z_{2+} = \left(Z_{12+} + Z_{23+} - Z_{13+}\right)/2 = \left(4.89 + 11.20 - 17.15\right)/2 = -0.53\%\%$$

$$Z_{3+} = \left(Z_{13+} + Z_{23+} - Z_{12+}\right)/2 = \left(17.15 + 11.20 - 4.89\right)/2 = 11.73$$

The impedance between HV and parallel connected LV and TV is

$$Z'_{1+} = Z_{1+} + \frac{Z_{2+}Z_{3+}}{Z_{2+} + Z_{3+}} = 5.42 + \frac{(-0.53) \times 11.73}{-0.53 + 11.73} = 4.86\%$$

The zero sequence impedance Z_{10} is estimated as

$$Z_{10} \approx 1.0 \times Z'_1 = 4.86\%$$

The reluctance of zero sequence flux path is almost the same as the one of the positive sequence flux path, so 1.0 coefficient number is used. The positive sequence impedance between HV and LV is $Z_{12+} = 4.89\%$. The zero sequence impedance between HV and LV is estimated as

$$Z_{120} \approx 1.0 \times Z_{12+} = 4.89\%$$

Z_{20} is estimated as

$$Z_{20} \approx Z_{120} - Z_{10} = 4.89 - 4.86 = 0.03\%$$

Z_{20} is tested -0.48%. Z_{30} is estimated as

$$Z_{30} \approx 1.0 \times Z_{3+} = 11.73\%$$

REFERENCES

1. Jim Fyvie, *Design Aspects Power Transformers*, Arima Publishing, UK, 2009.
2. Martin J. Heathcote, *The J&P Transformer Book,* 13th edition, Elsevier, Newnes, Amsterdam, et al, 2007.
3. Feng Cizhang, *Electric and Magnetic Fields*, 1979, People Education Publishing, Beijing, China, 1979.
4. S. V. Kulkarni, et al., *Transformer Engineering, Design and Practice*, Marcel Dekker, Inc., New York, 2004.
5. *IEEE Standard Test Code for Liquid-Immersed Distribution Power, and Regulating Transformers*, IEEE Std C57.12.90-2006, New York, USA, 2006.

6 Load Loss

When currents flow through windings, they produce I^2R or DC losses. Same currents generate a magnetic field inside of the tank which is called leakage flux. When this flux hits metallic parts, eddy currents and related loss in these metallic parts are generated; for winding, such loss is called eddy current loss. Besides winding's eddy current loss and I^2R loss, circulating current due to no or improper conductor transposition also generates loss. The eddy current losses in structural metallic parts such as core clamping structure and tank wall caused by the leakage flux are called tank stray losses. Transformer load loss is the sum of all these losses, and it exists in the form of heat. Understanding the load loss helps to reduce it. Knowing load loss distribution helps to control local heating which is important for large transformers. Each component of load loss is discussed in detail in the following sections.

6.1 I^2R LOSS

I^2R loss is loss produced by winding conductor DC resistance only. The specific loss, $P_{DC@75}$, in Watt/kg of copper and aluminum at 75°C, are

$$\frac{P_{DC@75}}{m} = 2.36 \times J^2 \text{ for copper}$$

$$\frac{P_{DC@75}}{m} = 13.3 \times J^2 \text{ for aluminum} \tag{6.1}$$

Where J is current density in A/mm^2, and m is conductor mass in kg. In the deduction of this equation, electrical copper conductivity of 58 m/Ω·mm^2 at 20°C and density of 8.89×10^{-3} kg/cm^3 are used. Typical current density in copper with class A insulation paper (105°C maximum temperature) such as cellulose paper is between 2 and 4 A/mm^2 [1,2]; the current density varies from 3.2 A/mm^2 for distribution transformers to 5.5 A/mm^2 for large transformers with forced cooling [3]. With Class H insulation paper (180°C maximum temperature) such as Nomex, current density can be around 7 A/mm^2; mobile power transformers often have such current density in order to make the small core and coil fit into a limited space. I^2R loss at other temperature, t, is

$$P_{DC} = \frac{234.5+t}{310} \times P_{DC@75} \text{ for copper}$$

$$P_{DC} = \frac{225+t}{300} \times P_{DC@75} \text{ for aluminium} \tag{6.2}$$

The minimum I^{2}R loss is achieved when $J_{LV} \approx J_{HV}$, where J_{LV} and J_{HV} are current densities of LV and HV windings respectively. The copper weight in kilogram, m, is

$$m = m_f \cdot N \cdot D \cdot \pi \cdot A \cdot d \times 10^{-6}; \ m = 8.89 \times 10^{-3} \times m_f \times l \times A \tag{6.3}$$

Where m_f is number of phases; N is number of turns; D is mean diameter in mm; l is wire length of one winding in meter; A is cross-sectional area of conductor in mm^2; d is density in kg/cm^3.

6.2 WINDING EDDY CURRENT LOSS

Winding eddy current loss can be accurately calculated by the Finite element analysis method through analyzing the leakage flux field the winding is in. The purpose of analytic study conducted here is to find relationships between the eddy current loss and other factors such as conductor strand dimension and winding height, to obtain a measure to reduce the eddy current loss. The leakage flux field in a typical two-winding transformer is shown in Figure 6.1. When leakage flux hits the winding conductor, a voltage is induced. This induced voltage produces eddy current in the conductor, the direction of which is 90° to the load current and the leakage flux, so the total current flowing in the conductor is $\sqrt{I_{eddy}^2 + I_{dc}^2}$. The peak value of axial leakage flux density is

$$B_m = \mu_0 \frac{\sqrt{2} \cdot I_1 N_1}{H_{wdg}} \cdot R_g = \mu_0 \frac{\sqrt{2} \cdot I_2 N_2}{H_{wdg}} \cdot R_g \tag{6.4}$$

Where $\mu_0 = 4\pi \times 10^{-7}$H/m, I_1 and I_2 are rms values of primary and secondary currents respectively in amperes; N_1 and N_2 are turn numbers of primary and secondary windings respectively, H_{wdg} is winding height in meter, R_g is Rogowski coefficient,

$$R_g = 1 - \frac{\Sigma a}{\pi \cdot H_{wdg}} \cdot \left(1 - e^{-\pi \frac{H_{wdg}}{\Sigma a}} \right), \Sigma a = a_1 + a_{12} + a_2 \text{ is total radial build of two windings}$$

plus the gap between them, as shown in Figure 6.1. As can be seen from Figure 6.1, axial flux is a major component. In the middle area of a winding, the leakage fluxes are almost of an axial nature; in the area of winding ends, there exists radial flux. In order to simplify the following deduction process of eddy current loss, it is assumed that only axial flux exists. This assumption gives satisfied engineering accuracy.

Figure 6.2 shows a conductor in a magnetic flux field, the flux in conductor area of $2x \cdot l$ is

$$\Phi_x = 2 \cdot x \cdot l \cdot B_x \tag{6.5}$$

Where l is conductor length and B_x is flux density. The rms value of voltage induced in this conductor area is

$$E_x = \sqrt{2}\pi \cdot f \cdot \Phi_x = 2\sqrt{2}\pi \cdot f \cdot x \cdot l \cdot B_x \tag{6.6}$$

Where f is frequency in Hz. The resistance of dx strip is

$$dR = \rho \frac{2 \cdot l}{b \cdot dx} \tag{6.7}$$

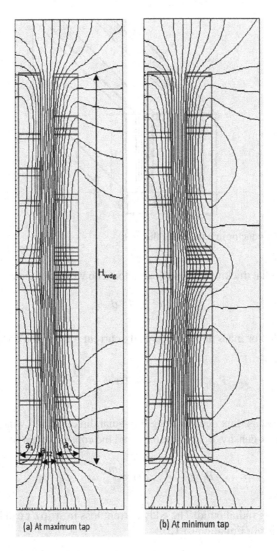

(a) At maximum tap   (b) At minimum tap

FIGURE 6.1 Leakage flux distribution.

Where b is the axial height of one conductor and ρ is conductor resistivity. Eddy current loss at the dx strip is

$$dP_{eddy} = \frac{E_x^2}{dR} = \frac{4\pi^2 \cdot f^2 \cdot l \cdot B_x^2 \cdot x^2 \cdot b}{\rho} \cdot dx \tag{6.8}$$

The eddy current loss at a whole cross-section of one conductor, P_{eddy}, is

$$P_{eddy} = \int_0^{a/2} dP_{eddy} = \frac{\pi^2 \cdot f^2 \cdot l \cdot B_x^2 \cdot a^3 \cdot b}{6\rho} \tag{6.9}$$

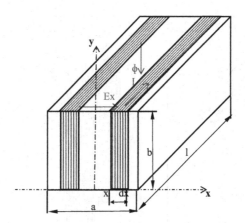

FIGURE 6.2 A conductor in a magnetic flux field.

Where a is the radial thickness of one conductor. The weight of the conductor, m, is

$$m = a \cdot b \cdot l \cdot d \tag{6.10}$$

Where d is conductor mass density. The eddy current loss of one conductor is then

$$P_{eddy} = \frac{\pi^2}{6 \cdot \rho \cdot d} \cdot m \cdot \left(a \cdot f \cdot B_x\right)^2 \tag{6.11}$$

The change of leakage flux density along the radial direction of a winding is shown in Figure 6.3. The flux density, B_x, at location x, and the copper weight of strip dx, dm, are

$$B_x = \frac{B_m}{c} \cdot x; \; dm = \frac{m}{c} \cdot dx \tag{6.12}$$

Where c is winding radial build. The eddy current loss in strip dx can be expressed in another way by using Equation (6.12).

$$dP_{eddy} = \frac{\pi^2}{6 \cdot \rho \cdot d} \cdot \frac{m}{c^3} \cdot \left(a \cdot f \cdot B_m\right)^2 \cdot x^2 \cdot dx \tag{6.13}$$

The eddy current loss of one section is:

$$P_{eddy} = \int_0^c dP_{eddy} = \frac{\pi^2}{18 \cdot \rho \cdot d} \cdot m \cdot \left(a \cdot f \cdot B_m\right)^2 \tag{6.14}$$

From Equation (6.4), the peak value of leakage flux, B_m, is

$$B_m = 4\pi\sqrt{2} \frac{I \cdot N}{H_{wdg}} \cdot R_g \times 10^{-7} = 17.77 \times \frac{I \cdot N}{H_{wdg}} \cdot R_g \times 10^{-7} \tag{6.15}$$

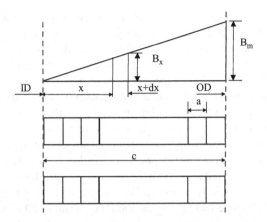

FIGURE 6.3 Leakage flux density variation along the radial build of inner winding.

Where H_{wdg} is winding axial height. The percentage of eddy current loss to I^2R is

$$P_{eddy}\% = \frac{\pi^2 \times 10^2}{18 \cdot \rho \cdot d \cdot I^2 \cdot R} \cdot m \cdot (a \cdot f \cdot B_m)^2$$

$$= \frac{17.77^2 \times 10^{-12} \times 50^2 \times \pi^2 \times a^2 \times A^2}{18 \cdot \rho^2} \cdot \left(\frac{f}{50}\right)^2 \cdot \left(\frac{N}{H_{wdg}}\right)^2 \cdot R_g^2 \qquad (6.16)$$

Where A is the cross-sectional area of a conductor of one turn. When there is more than one strand in one turn, the cross-sectional area of one turn is

$$A = a \cdot n_a \times b \cdot n_b \qquad (6.17)$$

Where n_a and n_b are numbers of the strand in radial and axial directions in one turn respectively. $P_{eddy}\%$ is then

$$P_{eddy}\% = \frac{17.77^2 \times 10^{-12} \times 50^2 \cdot \pi^2 \times a^4 \times n^2}{18 \cdot \rho^2} \cdot \left(\frac{f}{50}\right)^2 \cdot \left(\frac{b}{H_{wdg}}\right)^2 \cdot R_g^2 \qquad (6.18)$$

Where $n = N_{radial} \cdot n_a \times N_{axial} \cdot n_b$; $N = N_{radial} \times N_{axial}$. At 75°C, electrical copper resistivity is 2.097×10^{-8} Ω•m. The eddy current loss at 75°C is

$$P_{eddy@75}\% = 9.84 \times 10^{-4} \cdot a^4 \cdot n^2 \cdot \left(\frac{f}{50}\right)^2 \cdot \left(\frac{b}{H_{wdg}}\right)^2 \cdot R_g^2 \qquad (6.19)$$

Where a of the radial thickness of one conductor, b of axial height of one conductor and H_{wdg} of winding height are in mm. The percentage of eddy current loss of whole winding is same as the percentage eddy current loss of one section, since flux distributions in each section are the same under the assumption that only axial flux exists.

Also, the eddy current loss discussed here is caused only by winding flux itself; it doesn't include eddy current loss caused by other winding flux. From Equation (6.19), it can be seen that using a thin conductor can effectively reduce eddy current loss. Higher resistivity also makes the eddy current loss small; however, I^2R loss is increased. Transformers of low impedances have relatively low eddy current loss since low impedance means low leakage flux densities; their eddy current loss and winding hot spots are usually not a serious problem. On the other hand, transformers of high impedances have relatively high eddy current loss since their leakage flux densities are high; as a result, their eddy current losses and winding hot spot are sometimes a serious issue. In practice, well-designed units usually have eddy current loss of less than 5% in small units, 15% in large units of I^2R loss.

The eddy current loss in Equation (6.19) is average eddy current loss. As shown in Figure 6.3, the leakage flux density at the inner winding ID is zero, increases to maximum at the inner winding OD as well at the outer winding ID, then decreases to zero at the outer winding OD. The eddy current loss has the similar distribution pattern: the conductors at the OD of inner winding and the ID of outer winding which are adjacent to the main leakage flux channel have the highest eddy current loss, while the conductors at the ID of inner winding and the OD of outer winding have zero eddy current loss. The eddy current loss at other temperature, t, is

$$P_{eddy}\% = \frac{234.5+75}{234.5+t} \cdot P_{eddy@75}\% \text{ for cooper}$$

$$P_{eddy}\% = \frac{225+75}{225+t} \cdot P_{eddy@75}\% \text{ for aluminium}$$

(6.20)

The eddy current decreases exponentially from its maximum value at the conductor surface with depth into the conductor; the eddy current loss decreases in the similar way. This phenomenon is called skin effect; the depth where eddy current density decreases to 36.8% ($1/e$) of its original value at the conductor surface is called penetration depth, δ, which is [4]:

$$\delta = \frac{1}{\sqrt{(\pi \cdot f \cdot \mu \cdot \sigma)}}$$

(6.21)

where $\mu = \mu_r \mu_0$. The penetration depths of copper, aluminum and mild steel are listed in Table 6.1.

The bending of the flux lines at winding ends as shown in Figure 6.1 means that there is a radial component. This radial flux hits the width side of the winding conductors near the winding top and bottom and produces extra eddy current loss. Referring to Equation (6.11), it is

$$P_{radial\ eddy} = \frac{(2\pi f)^2 \cdot B_{radial}^2 \cdot b^2}{24 \cdot \rho} \left(\frac{m}{d}\right)$$

(6.22)

where $P_{radial\ eddy}$ is the eddy current loss produced by radial flux density B_{radial}. The result is that those conductors at the top of the winding are generally hotter than other conductors due to this extra loss; their temperatures are called winding hot spot

TABLE 6.1
Penetration Depth of Several Materials

	Copper	Aluminum	Core Steel	Magnetic Steel	Stainless Steel
Conductivity ($1/\Omega\cdot m$)	4.77×10^7 @ 75°C	2.88×10^7 @ 75°C	2.11×10^6	7×10^6 @ 20°C	1.136×10^6
Permeability	μ_0	μ_0	$4000\mu_0$	$100\mu_0$	μ_0
Penetration depth @50Hz, mm	10.3	13.2	0.77	2.69	66.8
Penetration depth @60Hz, mm	9.4	12.1	0.7	2.46	61.0

temperatures. The winding hot spot temperature rise above ambient has to be limited below a certain level in order to keep a certain transformer service life span, for example 80°C for a 65°C rise unit, 65°C for a 55°C rise unit.

The foregoing discussion concerns the eddy current loss caused by current flowing in the winding itself. In an idle winding which does not carry current but is in a magnetic flux field, eddy current loss is still produced in it; the detailed discussion of this is offered in [5]. The following is a list of measures to reduce eddy current loss and hot spot temperature rise.

- Using CTC (continuously transposed conductor) for high current carrying windings. The size of the conductor strand in CTC is usually smaller than the size of a regular single wire for the same application. From Equation (6.19), small strands of CTC reduce eddy current loss greatly comparing to multi-single wire per turn. In the case of helical windings having lower insulation levels, using netted or paperless CTC eliminates gradient drop across the paper wrap so that winding gradient is further reduced, and blockage of the oil duct by the paper bulge is avoided. These are the reasons netted CTC is used for LV windings of large units. When epoxy resin is applied on each strand, the strands in CTC are bonded together after heating, which improves the short-circuit strength of the windings greatly.

Along with the CTC advantage mentioned earlier, there are some disadvantages. First, overall size of CTC is larger than single wire, it requires a minimum inside diameter (ID) of the winding. For a smaller diameter, it is hard to wind CTC on the winding cylinder since CTC tends to spring off. Generally, the requested minimum winding ID of a CTC can be found by following this equation:

$$ID_{min} = (P \cdot N \cdot W) / \pi \quad (6.23)$$

where $P = 9$ to 18, N is number of strands of CTC and W is axial strand width. Secondly, paper wrap tends to bulge if CTC radial dimension is big. Such bulging restricts oil flow in the radial duct. The solution to this is to use several smaller CTCs to replace one big CTC to reduce the CTC radial dimension. Slightly increasing the radial oil duct size also helps.

- To reduce eddy current loss, as shown in Equation (6.19), the conductor surface which is hit by the leakage flux needs to be reduced, i.e., dimension a needs to be small. Caution is needed because from short-circuit strength considerations, there is a limitation on the minimum thickness. Also, when the width to thickness (b/a) ratio of the strand conductor is too large, it causes difficulties to wind a coil; the thin strand may tilt before the epoxy has cured, even under clamping pressure.
- Ampere-turn of HV and LV windings per unit height should be kept balanced as much as possible along the winding height, otherwise there is an extra radial flux and related loss. For example, HV winding, having taps within its body, the radial flux in the tap area as shown in Figure 6.1b, can cause excessive loss and temperature rise. Another important application of ampere-turn balance is to reduce axial short-circuit force exerting on winding conductors and clamping rings.
- Winding hot spot is generated by additional eddy current loss produced by radial flux as discussed earlier. Reducing the conductor width can reduce such eddy current loss, reducing the hot spot gradient. One practice to reduce radial flux eddy current loss is using a conductor of larger thickness and smaller width in top sections so that I^2R loss is approximately unchanged and the radial flux eddy current loss is reduced. Here it is assumed that the increase in axial flux eddy current loss due to larger thickness is compensated by the decrease in radial flux eddy current loss due to smaller width. The disadvantage is the amount of labor used in brazing the conductors, especially for CTC, and the risk of poor quality of the brazing. In selecting a conductor width without a serious hot spot issue in the absence of detailed analysis, for 50 Hz units, the maximum width that can be used is usually in the range of 12 to 14 mm. For 60 Hz units, it is between 10 and 12 mm. The relative height of LV and HV windings and each winding radial build also have an effect on eddy current losses and hot spot.
- In gapped core shunt reactors, there is considerable flux fringing between leg packets, resulting in an appreciable radial flux causing excessive loss in the reactor winding if the distance between the reactor winding and core is small, or if the conductor width is large.

6.3 CIRCULATING CURRENT LOSS

The leakage flux distribution in LV winding-gap-HV winding system is shown in Figure 6.4. When there is more than one conductor per turn, and these conductors are placed one by one radially, the leakage flux linked to each conductor is different. As the consequence, the induced voltages in each conductor are not the same. All of the conductors in one turn are crimped together at winding ends, a closed loop is made, the sum of induced voltages in the closed loop is not zero and it generates a current

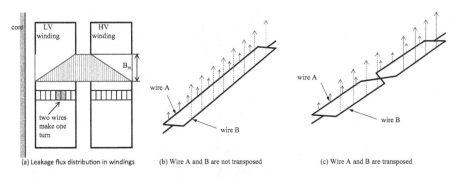

(a) Leakage flux distribution in windings   (b) Wire A and B are not transposed   (c) Wire A and B are transposed

FIGURE 6.4 Leakage flux distribution in windings and loop.

circulating in the loop and extra loss called circulating current loss. This loss depends much on the physical locations of conductors in the leakage flux field and conductor size.

To minimize the circulating current loss, transpositions are used. Transpositions make each conductor in one turn occupy the same axial physical location in the leakage flux field with the same conductor length, such that the induced voltages in every conductor are the same and cancel each other when they are crimped at the ends, so no circulating current is flowing in the loop. In practice, the circulating current is not entirely cancelled due to other design considerations or production imperfection; still, it is greatly reduced by the transposition. The transpositions in helical winding and disk winding will be discussed separately.

6.3.1 Transposition of Helical Winding

One example of commonly used transpositions scheme of helical winding is shown in Figure 6.5. in which one turn consisting of four conductors or CTCs needs three transpositions. Generally, one turn consisting of n conductors or CTCs needs $n - 1$ transpositions. As can be seen from Figure 6.5, each conductor occupies each location with the same length; the induced voltage in conductor is equal to each other, so after the conductors are crimped at the ends, the voltage in the loop is zero, so no circulating current flowing in the loop.

However, in a real practical manufacturing process, there is the chance either that transpositions are not made in ideal location or that some transpositions are missed. It is worth to know in such cases how much the loss will be. It can be said that a perfect transposition exists if

$$\sum_{P=1}^{m} \frac{n}{N} \cdot P \cdot (P - 1) = C \tag{6.24}$$

is the same for all conductors through whole winding. Where P is the radial position of one conductor in winding counting outwards from the zero leakage field point; m is the radial positions which one wire has in whole winding. N is the total number of turns per layer; n is the number of turns in position P; n/N is per unity turn in position P.

FIGURE 6.5 Three transpositions of four cables in helical winding.

Figure 6.6 shows the meaning of some parameters. If the transposition is not perfect, then C of each conductor is different, the extra loss caused by the circulating current, $Loss_{cir}$, is

$$\frac{Loss_{cir}}{Eddy_{wdg}} = \frac{45}{\left(5 \times M^2 - 1\right)} \left(\frac{Covered\ radial\ conductor\ width}{Bare\ radial\ conductor\ width}\right)^2$$

$$\left(\frac{1}{n_c} \sum_1^{n_c} \left(C_n - C_{average}\right)^2\right)$$

(6.25)

where n_c is number of conductors per turn and M is number of radial paths in parallel in a winding.

Example 6.1

A helical winding has 6 wires per turn and different transpositions. Its circulating current loss is listed in Table 6.2

It should be noted that the calculation per Equation (6.25) is based on the assumption that only axial leakage flux exists. Real transformer windings have both axial and radial flux. However, the axial component is much greater than the radial component, so Equation (6.25) gives a good estimation. For high current and low voltage helical winding, turns are made by either of several CTCs or group of single wires. With a high number of single wires, three transpositions are usually made in whole

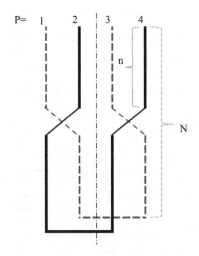

P= 1 2 3 4

n

N

FIGURE 6.6 Meaning of parameters in Equation (6.25).

TABLE 6.2

Transpositions and Circulating Current Loss of Helical Winding Having 6 Wires Per Turn

Number of layer = 1
Number of wires = 6
Covered wire thickness = 0.12
Bare wire thickness = 0.10

$$\sum_{P=1}^{m}\left(\frac{n}{N}\right)\cdot P\cdot(P-1)=C \quad C_n \quad (C_n - C_{average}) \quad (C_n - C_{average})^2$$

P 1 2 3   4 5 6
a b c   d e f

$n/N = 1/3$

Wire a
$$\frac{1\times0}{3}+\frac{3\times2}{3}+\frac{4\times3}{3}=6 \qquad 11\frac{2}{3} \qquad -5\frac{2}{3} \qquad 32.11$$

Wire b
$$\frac{2\times1}{3}+\frac{2\times1}{3}+\frac{5\times4}{3}=8 \qquad 11\frac{2}{3} \qquad -3\frac{2}{3} \qquad 13.44$$

Wire c
$$\frac{3\times2}{3}+\frac{1\times0}{3}+\frac{6\times5}{3}=12 \qquad 11\frac{2}{3} \qquad \frac{1}{3} \qquad 0.11$$

$n/N = 1/3$

Wire d
$$\frac{4\times3}{3}+\frac{6\times5}{3}+\frac{1\times0}{3}=14 \qquad 11\frac{2}{3} \qquad 2\frac{1}{3} \qquad 5.44$$

Wire e
$$\frac{5\times4}{3}+\frac{5\times4}{3}+\frac{2\times1}{3}=14 \qquad 11\frac{2}{3} \qquad 2\frac{1}{3} \qquad 5.44$$

$n/N = 1/3$

Wire f
$$\frac{6\times5}{3}+\frac{4\times3}{3}+\frac{3\times2}{3}=16 \qquad 11\frac{2}{3} \qquad 4\frac{1}{3} \qquad 18.78$$

$$\sum_{1}^{n_c}(C_n - C_{average})^2 = 75.32$$

$$\frac{1}{n_c}\sum_{1}^{n_c}(C_n - C_{average})^2 = 12.55$$

$$\frac{Loss_{cir}}{Loss_{eddy}} = \frac{45}{(5\times6^2-1)}\cdot\left(\frac{0.12}{0.1}\right)^2\times12.55 = 4.54$$

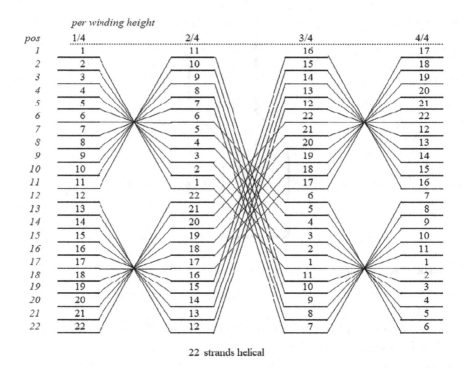

22 strands helical

FIGURE 6.7   Three transpositions of 22 strands in helical winding.

winding for easy manufacturing, Figure 6.7 lists the commonly used transpositions of 22 single wires per turn as an example. The conductor paper insulations in the transposition location is probably weakened or damaged due to bending the conductor; hence, the extra insulation wrap and pad are added to prevent short-circuit between the turns.

6.3.2 Transposition of Disk Winding

When the sums of square position number of each strand in each section get close to each other, the circulating current decreases to minimum. An example is shown in Figure 6.8.

For winding having line lead at its center, the top half and bottom half windings are parallel connected. The number of turns between these two halves, between the taps, must be the same; otherwise, a circulation current flows between them and extra load loss will result.

6.4 CIRCULATING CURRENT LOSS IN WINDING LEADS

Some regulating windings have tap leads made up of several conductors. Those conductors are brazed together to a certain location of the winding at one end, and crimped to a cable at another end, which make conducting loops. There are

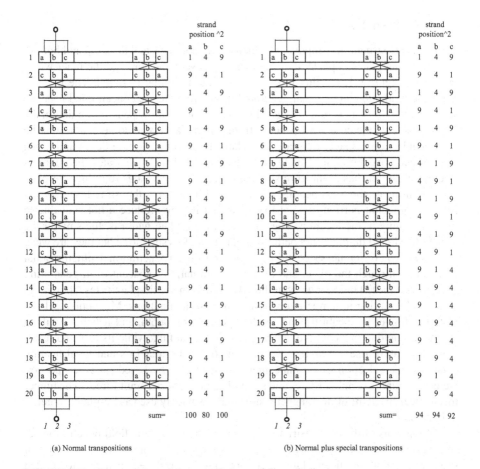

FIGURE 6.8   Transposition of three conductors of disc winding.

circulating currents in these loops, the values of which depend on the induced voltages and loop resistances. When units have high impedance and high radial leakage flux at regulating winding ends, the induced voltages are high. In order to reduce the circulating currents, rolling transposition is sometimes used. Caution must be exercised when applying the rolling transposition method.

6.5 LOSSES IN METALLIC STRUCTURE PARTS

When structural steels are subjected to a magnetic field, like winding copper conductor, an eddy current is induced in components such as tie-plate, yoke clamping frame and tank wall, eddy current loss is generated as the result, it is called stray loss. This loss is noticeable on large units, it sometimes results in local overheating issue besides extra load loss. Understanding the cause and taking preventive methods during design are important.

6.5.1 Tie-Plate Loss

Tie-plates are structural components which clamp core legs and hold top and bottom core clamping plates in place; they are located between the core and the inner winding. From Figure 6.1, the flux hitting on tie-plate is primarily radial. The induced magnetic field and eddy current generated in the tie-plate are concentrated near the surface if the tie-plate is magnetic steel because of its shallow penetration depth, the tie-plate shields the core steel beneath from this radial flux to some extent. On the other hand, stainless steel tie-plate has relatively deep penetration depth and much less surface concentration of eddy currents; the radial flux penetrates the stainless steel tie-plate, reaches at the core steel and heats it up. Methods to reduce tie-plate loss and mitigate the local heating are discussed here. First, this stray loss may not form a significant portion of total load loss, but it may produce a hot spot due to limited cooling in the location. The narrower the tie-plate width is, the smaller the loss. Fully slotted tie-plates are better for reduction of its stray loss, but they are weak mechanically. For stainless steel tie-plate, due to its deep penetration depth, the radial flux can go through the plate, and hitting on laminations adjacent to it, generating extra loss in these laminations even though the loss in the tie-plate is less than magnetic steel resulting in temperatures rising in these laminations. Magnetic steel tie-plate has higher permeability and shallow penetration depth, and it behaves like a magnetic shield to prevent core lamination from radial flux, while its loss and resultant temperature rise are higher. Keep in mind that besides loss and local heating concerns, the selection of tie-plate material must also be based on core and coil weight, winding clamping forces and strength against short-circuit forces. The higher mechanical strengths of the magnetic steel make is widely used as tie-plates material in large units.

Laminations of the core restrict eddy currents and magnetic field penetrations to some extent compared to equivalent size of solid metallic block, so the amount of stray loss is not significant, but a hot spot may be produced in it due to limited cooling. Stainless steel tie-plate makes things worse due to its long penetration depth which allows more radial leakage flux to go through and to hit the laminations. To reduce the eddy current loss and prevent local heating of these laminations, slitting the first step of the core is a simple and effective method. If the first step is less than about 12 mm thick, slit the second step.

6.5.2 Clamping Plate Loss

Clamping plates are used to clamp top and bottom yokes. They have a large surface area which offers efficient cooling, so hot spots seldom develop. However, when high current leads are in vicinity, the eddy current loss in the plate generated by the lead magnetic field can be large enough to raise local temperature high enough to produce gases. This is one issue needing attention for large units. The solution to that is to place both LV windings' start and finish leads in the same region, which is usually at the winding top; the current goes in and out at the same place, and in this way the sum of lead currents is zero, as are the lead's magnetic flux and the eddy current loss. Another practical approach is the insertion of stainless steel patch in the clamping plate where high current leads are nearby. The magnetic flux produces less eddy current loss in stainless steel than in magnetic steel.

The top and bottom clamping plates are mechanically connected through tie-plates to make the entire core clamping system strong. The system has to avoid any electrical loop(s) for current flowing. The clamping system has only one ground point with the same reason. The core is insulated to its clamping system and also has only one ground point.

6.5.3 Tank Wall Loss

The losses in the tank wall are caused by, first, fringing flux from the winding's ends and places such as tap section area of winding. These fluxes hit on the tank wall, producing eddy current loss, here called stray loss. Outer winding is sometimes a bit shorter than inner winding for other design considerations. If the outer winding is too short, more leakage fluxes hit on the tank wall, generating more loss. On the opposite, if the outer winding is taller than inner winding, it will force the fluxes back to the core, increasing stray loss in the tie-plate and core [6]. The tank wall stray loss forms a major part of total stray loss due to its large area, although the flux density hitting it is low. This stray loss can be reduced by increasing the distance between winding and tank wall, which results in a larger tank and more oil. It can also be reduced by magnetic or electromagnetic tank shunts. Above a certain MVA which the currents exceed a certain level or unit has higher than usual impedance; tank shunts should be provided to reduce the stray loss as well as the tank wall temperature rise.

The second source of magnetic flux which produces eddy current loss is from current carrying lead going through a hole in the tank cover or wall such as bushing leads. In practice when bushing current is higher than, say, 2000 A, non-magnetic material such as stainless steel inserts should be used to reduce local overheating.

The third source of magnetic flux is generated by current carrying leads which run parallel to the tank wall. When this flux hits the tank wall, extra losses are produced. Their contribution to total stray loss may be small, but this loss is so concentrated that it could cause local heating in the tank wall as well as oil gassing. Two types of distances from the lead to the tank wall need to be checked, and the larger one should be selected; one is dielectric clearance based on lead insulation level to the ground, another is thermal distance to avoid local tank wall overheating. For high current lead, its thermal distance is sometimes larger than its dielectric distance, and a larger tank is needed to meet the thermal distances request. Another alternative is using tanks shunts or non-magnetic steel inserts of the tank wall to mitigate local overheating without increasing the tank size.

The bolt joints between the cover and tank wall may become overheated and damaged if the current induced in the tank wall and the cover being forced to complete its path through the bolt is large. A solution to this is to connect the tank wall and cover by metallic strips (links) made of high-conductivity material, such as aluminum or copper, to provide a low-resistance alternative path for the induced currents to flow.

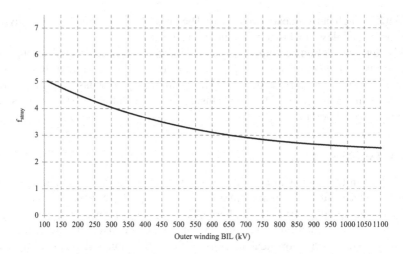

FIGURE 6.9   Relation of stray loss factor f_{stray} with outer winding BIL

The losses in tie-plate, clamping plate and tank wall make stray loss, and studying each of them individually is mainly for local overheating concerns. The stray loss is part of load loss which is usually guaranteed. During design, the stray loss has to be calculated, and one practical method is to use Equation (6.26) to estimate stray loss:

$$P_{stray} = f_{stray} \cdot MVA \cdot \left(\frac{Z}{100}\right) \cdot \left(\frac{f}{50}\right)^{0.8} \cdot \left(\frac{235+75}{235+t}\right) \tag{6.26}$$

where P_{stray} is stray loss, kW; MVA is transformer rating, MVA; Z is impedance, %; f is frequency, Hz; f_{stary} is a factor related to outer winding BIL, the tendency of which is shown in Figure 6.9. The calculated strays of most designs per Equation (6.26) are conservative; some units have higher-than-calculated stray loss due to high current leads being too close to the tank wall, which has no shunt.

6.6 SHUNTS

Shunts are an effective tool to reduce stray loss. Three types of materials used as shunts are copper, aluminum and core steel. Aluminum and copper shunts are sometimes called conductive shunts, while a core steel shunt is called a magnetic shunt.

Shunts reduce stray loss in the tank wall. More importantly, they prevent local overheating in the tank wall. The mild steel of the tank wall has short penetration depth, the loss in the tank wall concentrates in a shallow region below the surface. Stainless steel, on the other hand, has a much longer penetration depth. As a result, the local overheating in mild steel is more severe than in stainless steel. To mitigate local overheating, an option is using either stainless steel as the tank wall or stainless steel inserts in the tank wall.

6.6.1 SHIELD FROM WINDING LEAKAGE FLUX

The magnetic shunts are used to control the path of leakage fluxes by offering them a low-reluctance path. Take the core clamping plate shunt as example. When the clamping plates have a C shape, by placing such magnetic shunt on the horizontal flat part, the leakage fluxes which come out from windings' ends enter the shunts directly, in such a way that the amount of the flux in other places in tank is reduced. This results in less loss and heating of other structural components. Secondly, since the clamping shunts attract the leakage flux from winding ends, they make leakage flux contain more axial component, less radial component. This helps to reduce the loss produced by radial flux in winding ends areas, consequently, it helps to reduce the windings' hot spot temperature. Thirdly, since they shield the core clamping structure from the flux imping, the local heating in the structure is avoided. With adequate thickness, the loss in the shunt is low.

Two types of magnetic shunt exist: edgewise and widthwise. Edgewise is reported as having lower loss, but its manufacturing cost is high. Widthwise has loss not as low as edgewise, but its manufacturing cost is comparably lower. The thickness of shunt is decided by the flux density in it, which is in the range of 1.2~1.5 Tesla for grain-oriented steel shunt. When a unit encounters an overload, the effectiveness of shunt should be considered under the overload condition. To reduce the loss in clamp frame which has C shapes, not bar shape, shunts are laid on the flat surfaces of the frame facing coils. Sometimes, shunts cover the entire clamp frame length; it is said that this way, the flux from the three balanced phases will cancel each other out in the shunts. The leakage fluxes from three balanced phase coils cancel each other out when they meet regardless in oil or in shunt because they are 120 degrees apart. It is rather easier to make one shunt to cover the whole length of the clamp frame than to make three separated shunts for three phases. The study conducted in [7] shows that there is no advantage using shunt stacked on edge as compared with flat stacked. Other considerations, such as convenience, economics or possibly noise reduction, should dictate the choice.

6.6.2 SHIELD FROM HIGH CURRENT LEADS

The conductive shunt is used to shield the tank wall from magnetic flux generated by high current leads. The flux impinges the shield first, due to small penetration depths, most of the flux cannot go through the shield to hit on the tank wall. As a result, the tank wall is protected from being heated. Studies [4,8] show that in order to avoid the shunt themselves getting overheated, the minimum thickness of copper and aluminum shunts should be in the range of $(\pi/2)\cdot\delta$ to minimize its loss, where δ is penetration depth of the material. With 60 Hz application, copper has $\delta = 9.4$ mm, aluminum has $\delta = 12.1$ mm. This means that 15 mm and 20 mm thick of copper and aluminum shunts respectively have less chance to be overheated by their own loss. Another aspect needing attention is that the leakage flux repelled by a conductive shunt may hit other structural components, overheating them. For shielding a tank wall from a high-current magnetic field, the conductive shunts are better suited than magnetic shunts. This is because of the gaps between magnetic shunts, which reduce their effectiveness.

Practical methods to reduce local overheating in the tank wall caused by high current leads are summed up as follows. If low voltage phase current is over 4000 amperes, the tanks should be shielded from the lead. Instead of using shielding only in the vicinity of leads, a full shield should be used on the LV front to take advantage of stray loss reduction [6]. Loss in the tank wall caused by leads can be reduced by placing two or three phase leads as close as possible, since the magnetic fluxes from three phase leads cancel each other out. This is in fact an effective method for loss and local overheating reduction. Shunts are very effective in reducing the loss associated with one or two groups of leads from different phases, but not for three groups from three phases.

When the shunts are used for load loss reduction only, the cost of shunts plus assembly should be compared to capital of load loss.

REFERENCES

1. Jim Fyvie, *Design Aspects of Power Transformers*, Arima Publishing, UK, 2009
2. Martin J. Heathcote, *The J & P Transformer Book*, 13th Edition, Elsevier, Newnes, Amsterdam, et al, 2007.
3. R. Feinberg, *Modern Power Transformer Practice*, John Wiley & Sons, New York, 1979.
4. S.V. Kulkarni, et al., *Transformer Engineering, Design, Technology, and Diagnostics*, 2nd Edition, CRC Press/Taylor & Francis Group, Boca Raton, London, New York, 2013.
5. H. J. Kaul, *Stray-Current Losses in Stranded Windings of Transformers*, AIEE Transactions, June, 1957, pp 137–149, USA.
6. M. Waters, *The Short-Circuit Strength of Power Transformers*, MacDonald, London, 1966.
7. Robert M. Del Vecchio, et al, *Transformer Design Principles, with Applications to Core-form Power Transformers*, 2nd Edition, CRC Press/Taylor & Francis Group, Boca Raton, London, New York, 2010.
8. K. Karsai, et al, *Large Power Transformers*, Elsevier Science Publishing Company Inc., Amsterdam, Oxford, New York, Tokyo, 1987.

7 Cooling

7.1 BASIC KNOWLEDGE

The practical rating of a power transformer depends strongly on the heat generated by core, coil and the metallic parts hit by leakage flux, and on the cooling ability which takes the heat away. These two issues need detailed study and careful consideration when designing a new unit, or upgrading existing units to higher power rating levels. With the trends of using less material in the core and coil, and reduction in tank size and overall dimensions, the materials of core and coil are used more fully, and closer to their application limits, than ever before. So proper cooling of a unit nowadays is more important to keep that unit in service for its expected life span. This can be achieved by understanding heat generation and distribution inside a transformer; the application of new materials, for example thermal upgrade cellulose paper; and the application of more accurate calculating tools. Here at first, heat generation and distribution are discussed.

No-load loss generated in the core, load losses generated in windings and stray loss generated in structural parts, such as the tie-plate and tank wall, heat up the core, winding, structural parts, and their insulations respectively. All of the heat then transfers to the oil. The oil, besides as insulation medium, takes all of the heat and transfers it to the air through the radiator and tank, bringing down the temperatures of all these parts.

The heated oil becomes warm and has less density, as shown in Figure 7.1. The warm, lighter oil flows up along the windings, leaving the lower space for cold oil to fill. On the way up, the oil temperature keeps increasing because it is continuously heated up by windings along the way. The oil temperature reaches its maximum at the top of the windings. Then the oil enters into a cooler, such as a radiator, through its top valve, dissipates its heat to ambient through the radiator and is cooled. The cold oil has greater density so it is heavier, which makes it drop down to the bottom of radiator. From there the cold oil enters into the tank through the bottom valve, and restarts the cycle. As a result, the oil temperature along the winding height is not uniform. The oil is coldest at the bottom of the winding, the temperature gradually rises as the oil moves upward and reaches its maximum temperature at the top of windings, as shown in Figure 7.2.

The winding's losses heat up the oil by heating winding cables (paper insulated conductors) first to make them hotter than the adjacent oil. The difference in temperatures between the surrounding oil and the winding cable is defined as the winding gradient. When the conductor size and paper thickness of the cable are constant in an entire winding, the winding gradient is nearly constant at any part of winding because heat flux density (watt per heat dissipation surface area) at any part of winding is nearly constant. At the location of half of winding height, the temperature difference between winding and ambient cooling medium, such as ambient air, is defined as average winding temperature rise. From the winding bottom, the winding temperature rise increases nearly linearly and reach its maximum at the top of the

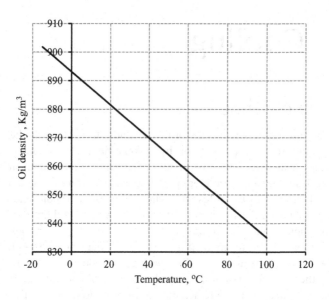

FIGURE 7.1   Oil density change with temperature.

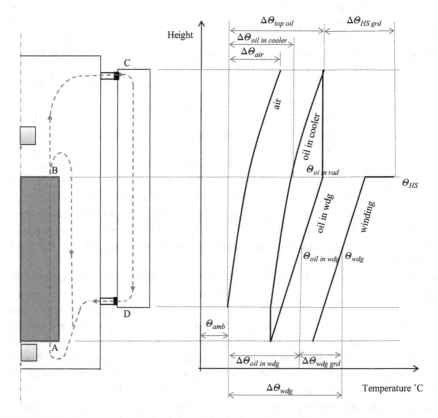

FIGURE 7.2   Temperatures, temperature rises and winding gradients.

winding. Extra heating occurs in the cables in the winding top and bottom, the result of extra eddy loss due to a larger amount of radial leakage flux hitting the top and bottom of winding. This extra heating further increases the temperature of the cables in top sections and creates a so-called hot spot; its temperature over ambient temperature is called hot spot temperature rise. Hot spot temperature is the determining factor to thermal aging of winding cable insulation, and the cable insulation aging is a determining factor to transformer service life.

7.1.1 AGING OF INSULATION PAPER

Cellulose materials used as winding cable insulation de-polymerize or age under high temperature. This aging (chemical reaction) process is accelerated by the presence of water and oxygen. The water exists as residual at the end of the drying process, or it comes through leaks in the tank, or as a by-product of the aging process. Oxygen can come into tank through leaks and the conservator. With the preservation system developed at present, the moisture and oxygen contribution to aging could be minimized.

De-polymerization breaks the long molecular chains of the cellulose material into short molecular chains. It reduces the material mechanical strength which is closely related to the length of the chains. The de-polymerization is indicated by degree of polymerization (DP). DP number decreases with an increase in the number of the short molecular chain, or a decrease in the tensile strength of cellulose paper. When either the DP number or the tensile strength is low enough, the paper as turn insulation becomes so brittle that a small movement of winding caused by short-circuit forces could tear down this paper, bring turn-to-turn dielectric failure and end the service life of the transformer. Note that even when the paper becomes brittle, it is still capable of taking normal electric stress due to oil impregnation and the barrier structure as long as the paper is not physically damaged.

A transformer's normal service life is determined by its insulation life. Based on numerous experiments and investigations, the aging rate of cellulose material follows the Arrhenius law as

$$Per\,Unit\,Life = Ae^{C-\frac{B}{T}} \tag{7.1}$$

where A, B and C are constants and T is absolute temperature, K. For a normal service life of 65°C rise unit, the normal hot spot temperature is $\Theta_{HS} = 30 + 80 = 110$°C. Through studies and investigations, it is found that for oil impregnated cellulose paper used in a power transformer, if the temperature is the only factor on aging, the effects of oxygen and water are not considered, Equation (7.1) becomes

$$Per\,Unit\,Life = 9.8 \times 10^{-18} \times e^{\frac{15000}{\Theta_{HS}+273}} \tag{7.2}$$

where Θ_{HS} is the winding hot spot temperature, °C. The per-unit life of unit with 110°C hot spot temperature is 1.0. When the load varies with time, the hot spot temperature varies with time. With variable ambient temperature with time, the hot spot temperature of a constant loading also varies with time through a delay by transformer time constant. The equivalent aging acceleration factor for a loading circle, F_{EQA}, is defined as

$$F_{EQA} = \frac{\sum_{n=1}^{N} F_{AA} \Delta t_n}{\sum_{n=1}^{N} \Delta t_n} \tag{7.3}$$

where F_{AA} is the aging acceleration factor

$$F_{AA} = e^{\left[\frac{15000}{383} - \frac{15000}{\Theta_{HS}+273}\right]} \tag{7.4}$$

N is total number of time intervals and Δt_n is time interval, hours. It is illustrated from Equation (7.4) that when the hot spot temperature is higher than 110°C, the aging of paper insulation is accelerated, making the unit service life span shorter; when the hot spot temperature is lower than 110°C, the aging slows down and the life span is extended compared to normal service life. The absolute life span of a transformer depends on the definition of end-of-life. One definition is 50% reduction in the tensile strength of winding cable paper insulation. It is found that this definition is too conservative. Sometimes transformers with 20% residual tensile strength operate well. Another definition of the end-of-life is the degree of polymerization in range of 100~250. Per IEEE Std C57.91-2011 [1], 200 is a good judge for power transformers, 100 is a good judge for distribution transformers. The new insulation papers usually have 1300 degrees of polymerization. The normal insulation life of a transformer is about 20 years. When the minimum life is 180,000 hours per IEEE Std C57.12.00 [2], the loss of life due to a temperature higher than 110°C is

$$\%loss\ of\ life = \frac{F_{EQA} \times 24 \times 100}{180000} \tag{7.5}$$

For a 65°C rise unit, if the normal hot spot temperature is 110°C, $F_{EQA} = 1.0$, and the loss of life is 0.013%. As a general rule, each 6-degree rise in winding hot spot temperature reduces the life of the insulation material by half.

7.1.2 Oil Thermal Behavior

The oil in the transformer tank serves two purposes. First, it works as electrical insulation. Besides its own excellent dielectric performance, the insulation papers and pressboards impregnated by oil have higher dielectric strengths than without. Another function of oil is as coolant; it takes the heat away from winding, core and structural parts, and it carries and then transfers such heat to the surrounding air through the tank wall and cooling system such as the radiator. Besides oil as insulation, power transformers usually have thermally upgraded Kraft paper as winding cable insulation; high-density pressboard as winding cylinder and spacers; medium-density pressboard as an insulation barrier; fiberglass, Nomex or ceramic dot matrix sheet as cooling duct in the core; Nomex as local insulations where the temperatures are higher and wood or pressboard as core step blocks.

7.1.3 Temperature Limits

The highest temperature and the average temperature of ambient cooling air over 24 hours is 40°C and 30°C respectively under the usual service condition, with non-thermally upgraded Kraft paper as winding cable insulation, the winding hot spot temperature is limited to 105°C, the hot spot rise is 65 (= 105 − 40)°C, the average winding rise is limited to 55°C. The average winding rise over ambient is the winding gradient over oil plus the average oil rise over ambient. With thermally upgraded Kraft paper as cable insulation, the winding hot spot rise is limited to 80 (= 120 − 40)°C, the average winding rise is limited to 65°C [2]. In order to achieve a normal service life span, the hot spot temperature of winding with thermally upgraded Kraft paper should be kept below 120°C. When the maximum temperature of ambient air exceeds 40°C, for example, 45°C, the winding hot spot temperature rise has to be reduced to 75°C in order to maintain the normal service life span.

Winding hot spot gradient is greater than average winding gradient, and the ratio of the two is called the winding hot spot factor. For small transformers, the hot spot factor is about 1.1; for medium and large transformers, the hot spot factor should be limited to 1.3. During short-circuit, which usually lasts 2 seconds for power transformers, the winding temperature is limited to 250°C for copper, and 200°C for aluminum, in this duration.

When the service site altitude is more than 1000 meters above sea level, the decrease of air density reduces the air cooling ability as well as its dielectric strength. Two methods are usually used to handle such situations. One, if the transformer is designed for usual service condition which is below 1000 meters in altitude, the transformer has to be de-rated in order to avoid overheating its windings caused by poor cooling ability of the air. Standard gives the de-rating factors for natural air cooling and forced air cooling [1]. Another method is to design the unit with extra cooling to compensate for the reduction of air cooling ability at high altitude. For water cooling units, due to the cooling ability being unaffected by the altitude, the units can operate without de-rating.

Under an overloading condition, higher winding hot spot and top oil temperatures than standard limits are allowed for short periods, as a sacrifice of life span. The winding cable insulation aging depends on the hot spot temperature and the duration of the overload; generally speaking, the shorter the duration is, the higher the hot spot temperature may be allowed with the same loss of life as the one for which the duration is longer but the hot spot temperature is lower.

Loading the transformer above nameplate rating requires careful consideration and experience, since it deals with loss of the service life of the unit. In addition to load capability of winding, the load capabilities of all other current-carrying components in the unit also play key roles. These components include bushings, tap changers, terminal boards, lead cables, current transformers, etc. Load beyond the limits of these components' capabilities could damage them. Another important issue is bubble evolution at high hot spot temperature. Studies show that moisture content in winding cable insulation is a key factor in the forming of bubbles, regardless of whether the oil preservation system is a nitrogen system or degassed system like a conservator system. At low moisture content levels, the bubble evolution temperature is virtually the same for both systems. At high moisture content levels, the effect of the gas content in the oil is noticeable. Aged transformers usually have ~2% moisture in the cable insulation paper, and the bubble evolution temperature is about 140°C. For a new transformer having 0.5% moisture content, the bubble evolution temperature is above 200°C. Most transformers have moisture levels in the range of 1~1.5%, hence it is prudent not to exceed 150°C hot spot temperature in any overloading case. The relation of bubble evolution temperature with moisture content and gas content is discussed in reference [1].

7.2 TEMPERATURE RISES OF OIL

As discussed earlier, oil circulation is driven by force which results from oil density change with temperature; this driving force is sometimes called gravitational buoyancy force. When the oil is heated to temperature Θ, its density, ρ, and specific gravity, γ, decrease as in Equation (7.6)

$$\rho = \rho_o \left(1 - \beta\Theta\right), \gamma = \gamma_0 \left(1 - \beta\Theta\right) \tag{7.6}$$

where ρ_o, γ_o are the values at a reference temperature. The heated oil has less specific gravity, then it flows upwards, and its original place is filled by the higher specific gravity cold oil. The oil circulating loop is shown in Figure 7.2. The cold oil entering into the winding bottom point A is heated and moves up to the winding top point B, then the hot oil enters into radiator point C, dissipates its heat to ambient through the radiator and descends to the bottom of the radiator. The cold oil leaves the radiator at point D, enters into the winding bottom point A, and starts the next circulation. The meanings of parameters used in Figure 7.2 and in the following are listed in Table 7.1. The average temperature of oil in radiator $\Theta_{oil\ in\ rad}$ is

$$\Theta_{\text{oil in rad}} = \Theta_{\text{amb}} + \Delta\Theta_{\text{oil in rad}} \tag{7.7}$$

If the unit is cooled by air circulation and natural oil flow, the average oil rise in the radiator $\Delta\Theta_{oil\ in\ rad}$ is slightly different from the average oil rise in the winding $\Delta\Theta_{oil\ in\ wdg}$. To simplify the thermal calculations, it is made

$$\Delta\Theta_{oil\ in\ rad} = \Delta\Theta_{oil\ in\ wdg} \tag{7.8}$$

If the unit is cooled by forced oil circulation, then

$$\Delta\Theta_{\text{oil in rad}} < \Delta\Theta_{\text{oil in wdg}} \tag{7.9}$$

Test results and researches show that for natural oil flow, the average oil rise in winding over ambient, $\Delta\Theta_{oil\ in\ wdg}$ is

$$\Delta\Theta_{oil\ in\ wdg} = k \cdot \left(\frac{W}{S}\right)^{0.8} \tag{7.10}$$

TABLE 7.1
Meaning of Each Parameter Used

Physical Meaning	Symbol	Relation
Top oil rise	$\Delta\Theta_{top\ oil}$	
Bottom oil rise	$\Delta\Theta_{bottom\ oil}$	
Average oil rise in winding	$\Delta\Theta_{oil\ in\ wdg}$	$\Delta\Theta_{oil\ in\ wdg} = (\Delta\Theta_{top\ oil} + \Delta\Theta_{bottom\ oil})/2$ $\Delta\Theta_{oil\ in\ wdg} = k \cdot \left(\dfrac{W}{S}\right)^{0.8}$
Average oil rise in radiator	$\Delta\Theta_{oil\ in\ rad}$	$\Delta\Theta_{oil\ in\ rad} = \Delta\Theta_{oil\ in\ wdg}$ For air circulation and natural oil flow
Difference between inlet and outlet of radiator	$\Delta\Theta_{oil}$	$\Delta\Theta_{oil} = \Delta\Theta_{top\ oil} - \Delta\Theta_{bottom\ oil}$
Winding gradient	$\Delta\Theta_{wdg\ grd}$	$\Delta\Theta_{wdg\ grd} = \Delta\Theta_{p} + \Delta\Theta_{s}$ $\Delta\Theta_{p}$ and $\Delta\Theta_{s}$ are temperature drops in paper and the paper surface to oil respectively
Winding hot spot gradient	$\Delta\Theta_{HS\ grd}$	$\Delta\Theta_{HS\ grd} = K_{HS} \cdot \Delta\Theta_{wdggrd};\ K_{HS}=1.1\sim1.3$
Winding hot spot rise	$\Delta\Theta_{HS}$	$\Delta\Theta_{HS} = \Delta\Theta_{HS\ grd} + \Delta\Theta_{top\ oil}$
Average winding rise	$\Delta\Theta_{wdg}$	$\Delta\Theta_{wdg} = \Delta\Theta_{wdg\ grd} + \Delta\Theta_{oil\ in\ wdg}$
Ambient temperature	Θ_{amb}	
Average oil temperature in radiator	$\Theta_{oil\ in\ rad}$	$\Theta_{oil\ in\ rad} = \Delta\Theta_{oil\ in\ rad} + \Theta_{amb}$
Average winding temperature	Θ_{wdg}	$\Theta_{wdg} = \Delta\Theta_{wdg} + \Theta_{amb}$
Winding hot spot temperature	Θ_{HS}	$\Theta_{HS} = \Delta\Theta_{HS} + \Theta_{amb}$

where W is total loss, which is load loss plus no-load loss, in Watts; S is effective heat dissipating surface of the radiator and tank wall plus cover, in m²; k is a coefficient. As can be seen from Equation (7.10), the average oil temperature rise is dependent on the losses and the heat dissipation surface.

When designing a transformer, efforts are made to keep altitude of the radiator higher. By doing so, the average oil temperature rise can be raised without raising top oil rise ($\Delta\Theta_{top\ oil} = \Delta\Theta_{oil\ in\ wdg} + \frac{1}{2}\Delta\Theta_{oil}$) as well as hot spot temperature rise; as a result, the capacity and the cost of cooling equipment may be reduced. The relation between $\Delta\Theta_{oil}$ and amount of heat P (loss) to be transferred is

$$P = c \cdot \Phi \cdot \Delta\Theta_{oil} \tag{7.11}$$

where c is specific heat and Φ is mass of flowing oil. With a certain amount of heat P to be transferred and a small $\Delta\Theta_{oil}$ being required, Φ has to be increased. A higher Φ can be achieved by increasing the driving force to make oil flow faster. The studies show that the driving force is related to the thermal head, which is the difference between radiator center altitude and winding center altitude. The test results show that the difference between top oil rise and average oil rise related to (X/H) has the trend as shown in Figure 7.3, where H is radiator height and $X = X_1 - X_2$; X_1 and X_2 are dimensions shown in Figure 7.4. Here, X is a dimension related to the thermal head but not *the* thermal head; X is used just for convenience. By mounting the

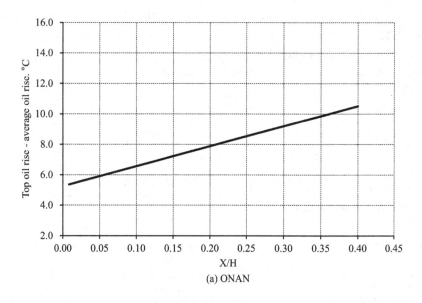

(a) ONAN

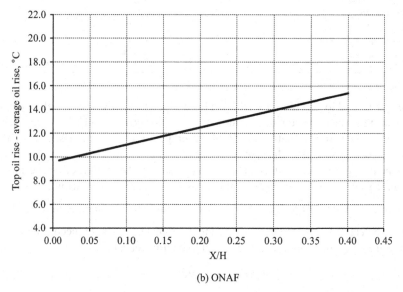

(b) ONAF

FIGURE 7.3 Relations of top oil rise-average oil rise with *X/H*.

radiator higher, the thermal head increases, the value of *X* decreases, and as a result the difference between the top oil rise and average oil rise decreases. However, the radiator cannot be mounted too high, otherwise, the oil leaving the radiator bottom does not reach the winding bottom and flow up through the windings; instead it bypasses the windings, flowing up to the top through the gap between the winding and the tank wall.

Choking factors such as the vertical ducts between winding and cylinder or wrap, horizontal ducts between sections of winding, bottom and top oil ducts in yoke pads

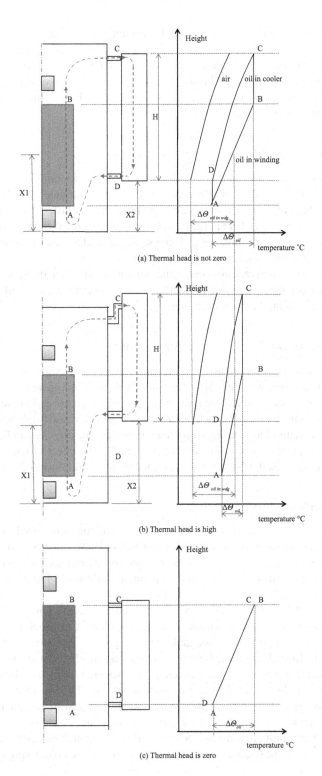

FIGURE 7.4 Effect of thermal head on oil rise.

through which oil flows in and out from the winding, the flange size of the radiator, etc., also impact mass of oil flow. Some of these are discussed in Section 7.4.2.

The oil circulates either naturally by its own gravitational buoyancy force or by exerting force from the pump. The cooling medium is either air or water. In the case of air as the external cooling medium, the air can blow radiator either naturally or be forced by fans. Each combination is discussed next.

7.2.1 NATURAL OIL FLOW, NATURAL AIR COOLING

Figure 7.4 shows the effect of radiator altitude referring to windings on oil and winding temperature. When the radiator is mounted at higher altitude, as in case (b), the gravitational buoyancy force is higher, and the oil circulates faster. As the result, with the same amount losses or heat in the tank, the oil temperature difference between the top and bottom of the winding, $\Delta\Theta_{oil}$, decreases compared to case (a), while the average oil rises over ambient, $\Delta\Theta_{oil\,in\,wdg}$, is unchanged.

As an extreme, when radiator center and winding center have the same altitude as shown in case (c), there is no gravitational buoyancy force to circulate oil. The radiator has lost its cooling function significantly.

7.2.2 NATURAL OIL FLOW, FORCED AIR COOLING

Due to high velocity of forced air, heat exchange on the air side increased. Compared to natural air cooling, with the same amount of loss to be dissipated, the oil will be cooled further, the oil temperature rise, $\Delta\Theta_{oil\,in\,wdg}$, will decrease and branch C-D will be somewhat more curvedly bended, as shown in Figure 7.4 case (a). It is said that by changing from natural air cooling to forced air cooling, the cooling can be improved about 2.6 times at the same ambient temperature [3]. The effect of radiator altitude on oil rises is same as the ONAN case since the nature of oil flow doesn't change.

7.2.3 FORCED OIL FLOW COOLING

For some larger transformers, a greater number of radiators is required to cool down the amount of heat from windings and other parts. These radiators are usually of large sizes. Sometimes, physical space or other considerations limit these. In this situation, changing the cooling method from natural oil flow to forced oil flow is an option.

Two types of forced oil cooling are often used at present. One is non-directed forced oil cooling (such as oil forced air forced, or OFAF), in which most of the oil at tank bottom is pumped up to the tank top through the gaps between the winding and tank wall. In such case, the heat transfer process inside the winding is affected little. The heat flux per unit transfer area (e.g. watt/square millimeter) allowed for the winding must not be higher than that for natural oil flow. The role of the pump is to mix the hot oil ascending from inside windings along path a, with the cold oil pumped up along the tank wall of path b, at the top of the winding, as shown in Figure 7.5a. The temperature of oil entering the radiator will be lower than the temperature of oil at the top of the winding. So using the temperature rise of the oil entering the radiator

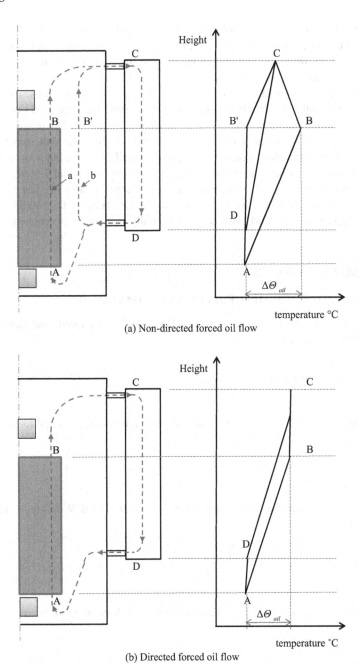

(a) Non-directed forced oil flow

(b) Directed forced oil flow

FIGURE 7.5 Forced oil and forced air cooling.

to decide loading capability is not correct. Both the oil flow velocity in the winding ducts and the temperature difference between the winding top and bottom changes little compared to natural oil flow.

Another type of forced oil cooling is called directed forced oil flow cooling (such as oil directed air forced, or ODAF). The oil is pumped to the winding bottom and is forced to flow up through the winding. In this case, the velocity of oil in the winding is increased; more heat is taken away from the winding. As a result, the temperature difference between top and bottom winding is reduced to about 2 degrees, and the heat flux per unit transfer area can be designed higher than one for natural oil flow without exceeding average oil temperature rise, as shown in Figure 7.5b. There are economic and technical limits imposed on increasing the heat flux per unit transfer area. Compared to non-directed forced oil cooling, a smaller-size radiator or cooler for directed forced oil cooling can be used. Under the directed forced oil flow, the winding surface temperature gradient decreases to such a level at oil velocity of a few dm/second that it is not worth it to increase the oil velocity further, because this would unnecessarily increase pumping work without proportionally reducing temperature rises.

7.3 LOADING CAPACITY

7.3.1 Ultimate Temperature Rises under Different Load

Ultimate temperature rises under an actual load following rated load can be estimated. Top oil rise at actual load is

$$\Delta\Theta_{top\,oil\,ultimate} = \Delta\Theta_{top\,oil\,rated} \cdot \left(\frac{1 + L_{load} \cdot R_{load}^2}{1 + L_{load}} \right)^n \qquad (7.12)$$

where $\Delta\Theta_{top\,oil\,rated}$ is top oil rise at rated load. Average oil rise at actual load is

$$\Delta\Theta_{oil\,in\,wdg\,ultimate} = \Delta\Theta_{oil\,in\,wdg\,rated} \cdot \left(\frac{1 + L_{load} \cdot R_{load}^2}{1 + L_{load}} \right)^n \qquad (7.13)$$

where $\Delta\Theta_{oil\,in\,wdg\,rated}$ is average oil rise in winding at rated load. Winding hot spot rise at actual load is

$$\Delta\Theta_{HS\,ultimate} = \Delta\Theta_{top\,oil\,rated} \cdot \left(\frac{1 + L_{load} \cdot R_{load}^2}{1 + L_{load}} \right)^n + K_{HS} \cdot \Delta\Theta_{wdg\,grd\,rated} \cdot R_{load}^{2m} \qquad (7.14)$$

where $\Delta\Theta_{wdg\,grd\,rated}$ is winding gradient at rated load. Average winding rise at actual load is

$$\Delta\Theta_{w\,ultimate} = \Delta\Theta_{oil\,in\,wdg\,rated} \cdot \left(\frac{1 + L_{load} \cdot R_{load}^2}{1 + L_{load}} \right)^n + \Delta\Theta_{wdg\,grd\,rated} \cdot R_{load}^{2m} \qquad (7.15)$$

$$R_{load} = \frac{actual\,load}{max\,continuous\,rating}; \; L_{load} = \frac{load\,loss\,at\,rated\,load}{no-load\,loss\,at\,rated\,voltage} \qquad (7.16)$$

K_{HS} is hot spot factor, n is exponential power of oil temperature rise, m is one-half of exponential power of winding temperature rise, the suggested values are $n = 0.8$ for ONAN, $0.9{\sim}1.0$ for ONAF and forced oil, $m = 0.8$ for ONAN and ONAF, $0.8{\sim}1.0$ for forced oil [1,4]. These suggested vales are conservative. IEEE standard C57.119 [5] introduces the test measurement procedures of these exponents. The measurements on the real units give $n \approx 0.75$, $m \approx 0.65$ at ONAF/ONAF rating.

7.3.2 INSTANT TEMPERATURE RISES

Transformer cooling and warming follows an exponential law. For example, the top oil temperature rise with time after a step load, $\Delta\Theta_{top\ oil}(t)$, can be expressed as

$$\Delta\Theta_{top\ oil}(t) = \Delta\Theta_{topoil\ initial} + \left(\Delta\Theta_{topoil\ ultimate} - \Delta\Theta_{topoil\ initial}\right)\left(1 - e^{-t/\tau_{TO}}\right) \quad (7.17)$$

where $\Delta\Theta_{topoil,ultimate}$ is ultimate top oil rise corresponding to the step load applied, $\Delta\Theta_{topoil,initial}$ is the initial top oil rise at $t = 0$ moment. Both $\Delta\Theta_{topoil,ultimate}$ and $\Delta\Theta_{topoil,initial}$ can be estimated by Equation (7.12). τ_{TO} is the oil time constant, which can be calculated as follows:

$$\tau_{TO} = \tau_{top\ oil\ rated}\ \frac{\left(\dfrac{\Delta\Theta_{top\ oil\ ultimate}}{\Delta\Theta_{top\ oil\ rated}}\right) - \left(\dfrac{\Delta\Theta_{top\ oil\ initial}}{\Delta\Theta_{top\ oil\ rated}}\right)}{\left(\dfrac{\Delta\Theta_{top\ oil\ ultimate}}{\Delta\Theta_{top\ oil\ rated}}\right)^{\frac{1}{n}} - \left(\dfrac{\Delta\Theta_{top\ oil\ inital}}{\Delta\Theta_{top\ oil\ rated}}\right)^{\frac{1}{n}}} \quad (7.18)$$

$$\tau_{top\ oil\ rated} = \frac{C \cdot \Delta\Theta_{top\ oil\ rated}}{P_{T,R}}$$

where $\Delta\Theta_{top\ oil\ rated}$ is the top oil rise at rated load, °C. C is thermal capacity which can be calculated per reference [1], while $P_{T,R}$ is the total loss at rated load, W. Winding hot spot gradient with time after a step load follows the same exponential pattern:

$$\Delta\Theta_{HS\ grd}(t) = \Delta\Theta_{HS\ grd,initial} + \left(\Delta\Theta_{HS\ grd,ultimate} - \Delta\Theta_{HS\ grd,initial}\right)\left(1 - e^{-t/\tau_w}\right) \quad (7.19)$$

where $\Delta\Theta_{HS\ grd}(t)$ is winding hot spot gradient at time t, °C. $\Delta\Theta_{HS\ grd,initial}$ is initial winding hot spot gradient at $t = 0$, $\Delta\Theta_{HS\ grd,ultimate}$ is ultimate winding hot spot gradient after the step load, °C. Both $\Delta\Theta_{HS\ grd,initial}$ and $\Delta\Theta_{HS\ grd,ultimate}$ can be estimated by Equation (7.14). τ_w is the time constant of the winding gradient, which usually is in the range between 3 and 15 minutes. The time needed for oil rise to be stable from one load to another is usually between 4 and 8 hours for natural oil and air flow, and between 1 and 3 hours for natural oil flow and forced air flow.

Example 7.1

For a 100/134/168 MVA unit, the load loss at 100 MVA and 85°C is 218.8 kW and the no-load loss at 100% voltage is 46.45 kW at 85°C. At top rating of 168 MVA, the top oil rise is 50.5°C and the average oil rise is 40.5°C. HV coil gradient is 18.5°C, the hot spot factor is 1.15. The hot spot gradient is 18.5 × 1.15 = 21.3°C. The unit will be overloaded to 120% of 168 MVA for 10 hours at 35°C ambient temperature. The temperature rises at the overload have to be known during the design stage. The load loss at 168 MVA is

$$Load\ loss\ at\ 168\ MVA = (168/100)^2 \times 218.8 = 617.5\ kW$$

$$R_{load} = 1.2; L_{load} = 617.5/46.45 = 13.29$$

The final temperature rises are stable to ultimate values during 10 hours of the overload. The final top oil rise at 120% overload is

$$\Delta\Theta_{top\ oil\ ultimate} = 50.5 \times \left(\frac{1+13.29 \times 1.2^2}{1+13.29} \right)^{0.9} = 68.8°C$$

The final top oil temperature is $\Theta_{top\ oil}$ = 68.8 + 35 = 103.8°C.
The final winding hot spot rise is $\Delta\Theta_{HS}$ = 68.8 + 21.3 × 1. $2^{2 \times 0.8}$ = 97.3 ° C
The final winding hot spot temperature is $\Theta_{hot\ spot}$ = 97.3 + 35 = 132.3 °C.
The loss of life calculation is based on Equations (7.4) and (7.5). The details are listed in Figure 7.6 which is output of a calculation program.

The calculation of loss of life

time	Δt (hours)	pu load	ambient (°C)	top oil rise $\Delta\Theta_{top\ oil\ initial}$ (°C)	top oil rise $\Delta\Theta_{top\ oil\ final}$ (°C)	HS rise $\Delta\Theta_{HS\ final}$ (°C)	F_{AA}	Hot Spot temperature (°C)	Top Oil temperature (°C)
1	1	1	35.0	50.5	50.5	71.8	0.719	106.8	85.5
2	1	1	35.0	50.5	50.5	71.8	0.719	106.8	85.5
3	1	1	35.0	50.5	50.5	71.8	0.719	106.8	85.5
4	1	1	35.0	50.5	50.5	71.8	0.719	106.8	85.5
5	1	1	35.0	50.5	50.5	71.8	0.719	106.8	85.5
6	1	1	35.0	50.5	50.5	71.8	0.719	106.8	85.5
7	1	1.2	35.0	50.5	56.1	84.6	2.606	119.6	91.1
8	1	1.2	35.0	56.1	60.0	88.5	3.785	123.5	95.0
9	1	1.2	35.0	60.0	62.6	91.2	4.882	126.2	97.6
10	1	1.2	35.0	62.6	64.5	93.0	5.812	128.0	99.5
11	1	1.2	35.0	64.5	65.8	94.3	6.552	129.3	100.8
12	1	1.2	35.0	65.8	66.7	95.2	7.117	130.2	101.7
13	1	1.2	35.0	66.7	67.3	95.8	7.535	130.8	102.3
14	1	1.2	35.0	67.3	67.7	96.3	7.839	131.3	102.7
15	1	1.2	35.0	67.7	68.0	96.6	8.056	131.6	103.0
16	1	1.2	35.0	68.0	68.2	96.8	8.210	131.8	103.2
17	1	1	35.0	68.2	62.8	84.1	2.482	119.1	97.8
18	1	1	35.0	62.8	59.0	80.3	1.711	115.3	94.0
19	1	1	35.0	59.0	56.4	77.7	1.317	112.7	91.4
20	1	1	35.0	56.4	54.6	75.9	1.096	110.9	89.6
21	1	1	35.0	54.6	53.3	74.6	0.964	109.6	88.3
22	1	1	35.0	53.3	52.5	73.8	0.881	108.8	87.5
23	1	1	35.0	52.5	51.9	73.2	0.828	108.2	86.9
24	1	1	35.0	51.9	51.4	72.7	0.793	107.7	86.4

$$\sigma_{n=1}^{N} F_{AA}\Delta t = 76.779$$
$$F_{EQA} = 3.199$$

	Max	Max
% loss of life = **0.043**	**131.8**	**103.2**

FIGURE 7.6 Output of overloading calculation.

7.3.3 WINDING HOT SPOT RISE

The direct measurements of winding hot spot rise show that, first, the hot spot location is not always at the highest loss density location, which is often at the top section of winding. The hot spot location depends on both local heat generation and cooling condition. Secondly, hot spot factor (hot spot gradient/winding gradient) increases with an increase of load. At present, the common practice takes the hot spot factor as a constant unchanged with load. Finally, the hot spot location changes with different loads. The present common practices assume the hot spot is located at the winding top regardless of the loads. The findings from reference [6] indicate more understandings are needed on characteristics of the winding hot spot.

7.4 COOLING OF WINDING

Two factors decide the temperature of a cable in winding, losses generated in this cable conductor and its cooling condition.

7.4.1 LOSSES GENERATED IN WINDING CABLE

When a transformer is loaded, the winding's cables generate losses. The losses generated by each cable in the same winding differ from each other and depend on the cable location. Three facts cause this phenomenon. First, the axial leakage flux starts from zero at the inside diameter of the inner winding, increases to peak level at the outside diameter of the inner winding, keeps the same level in the gap between the two windings, then decreases to zero at the outside diameter of the outer winding, as discussed in Chapter 6. Consequently, the cables at the outside diameter of the inner winding and inside diameter of the outer winding see this peak level of the flux, producing maximum axial eddy current loss. The cables at the inside diameter of the inner winding and outside diameter of the outer winding see minimum axial flux, producing the minimum axial eddy current loss.

The second cause of unevenly distributed loss is the radial leakage flux. The radial leakage flux may have negligible magnitude in the middle of a winding where there is no big gap; it is significant at the ends of the winding and varies with cable location. Figure 7.7 gives an example of radial flux at a winding top section. The same phenomenon happens when there is a big gap such as compensation gap in LV winding, or tap gap in HV winding. This radial flux, which is the lowest at locations adjacent to the main leakage flux channel and increases towards another edge of the section, produces the lowest radial eddy current loss in the cable next to the main leakage flux channel, the highest same loss in the cable at another edge of the section. The total eddy loss is the sum of axial and radial eddy current losses, the location of the highest such loss changes with different designs.

The third cause is oil, in which winding cables immerse, and which has different temperatures at different altitudes. At the bottom of a winding, for example, the oil has the lowest temperature, and the cables in this location have lowest resistance and lowest I^2R loss. At the top of the winding, the oil temperature reaches its maximum, and the cables in this location have the highest resistance and the highest I^2R loss. Although the eddy current loss is inversely proportion to temperature, due to its small

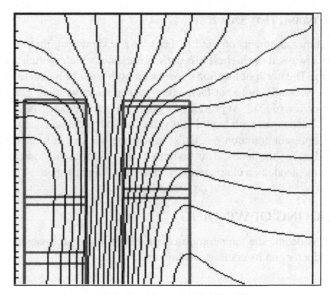

(a) Leakage flux at winding tops

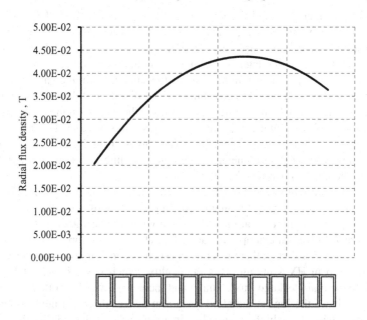

(b) Radial flux density distribution at top section of outer winding

FIGURE 7.7 Radial flux at the winding end.

portion in total winding loss, the overall winding loss will increase with temperature increase, which means that cables at the top of the winding generate more losses than cables at the bottom of the winding, even though both have the same current.

7.4.2 WINDING COOLING CONDITION

Oil circulation occurs when a transformer is energized and its thermal head is not zero. Cold oil at the winding bottom coming from radiator is heated by winding conductors and rises up towards the winding top. Hot oil at the winding top then goes to the radiator, through which the heat is dissipated to air. To make enough oil flow through the windings to carry enough heat, it is necessary to make the vertical oil duct next to the winding wide enough to reduce flow resistance. This cooling requirement conflicts with the dielectric requirement; to achieve high dielectric strength, this duct needs to be as thin as possible. Balancing these two factors, ducts with a thickness of no less than 6 mm and no more than 12 mm are usually used.

A study shows that a boundary oil layer exists adjacent to the winding surface, and the thickness of the boundary layer is about 6.5 mm [3]. 90% oil flow passes in this layer, with maximum velocity near the winding surface. Any duct size smaller than the thickness of the boundary layer increases the oil flow resistance and decreases the flowing mass, consequently increasing the temperature difference between the oil entering and the oil leaving the winding. A minimum duct thickness of 6.5 mm seems to be reasonable, but more studies are needed on the thermal behavior of the smaller duct. On the other hand, a duct size much larger than the thickness of the boundary layer does not significantly improve the flow condition.

When units have higher losses, the oil ducts next to windings and in yoke pads have to be wide enough to deliver enough oil to the windings. Generally speaking, for ONAN and ONFA cooling mode, when total loss is higher than 300 kW, this consideration is needed.

7.4.2.1 Winding Gradient with Natural Oil Flow

A larger cooling surface of a winding section brings better cooling to its conductors to keep their temperatures low. This requires fewer numbers of radial spacer per circle and narrower spacer width. However, the numbers of spacer per circle and spacer width have also to satisfy the requirement of short-circuit strength. Further, whether oil flows in horizontal ducts in winding is also a key factor. It depends on the sizes of horizontal duct d_h, the winding radial build RB as shown in Figure 7.8a and value of heat flux per unit transfer area, q. The main concern here is on d_h, since in most designs the vertical duct d_{v1} and d_{v2}, as shown in Figure 7.8a, is larger than d_h. If d_h is large enough, RB is small enough and q is also small, oil will flow through the horizontal duct d_h. When there is an oil flow in the horizontal ducts, convection plays a role in the heat transfer, and horizontal surfaces of the winding sections are convection surfaces. If d_h is small, RB is large and q is also large, the oil may not flow through horizontal ducts d_h. When there is no oil flow in the horizontal ducts, and the horizontal surfaces of the winding sections are not the convection surfaces, the heat is then transferred by conduction instead of convection. Larger transformers always have large RB due to large cable size, small d_h due to limited space and high value of q due to high winding loss. In order to make the oil flow through horizontal duct d_h, a so-called oil flow guide washer, as shown in Figure 7.8b, is used to guide oil flow. With the oil flow guide washer, the oil flowing through winding horizontal ducts is

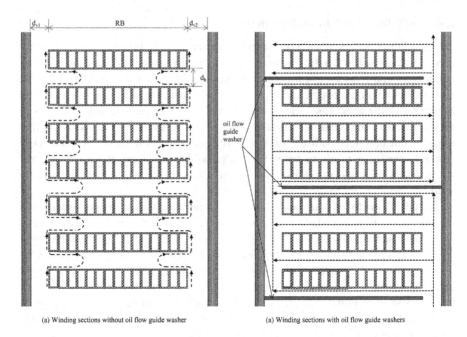

(a) Winding sections without oil flow guide washer   (a) Winding sections with oil flow guide washers

FIGURE 7.8   Oil flow in winding ducts.

achieved, and the heat from winding is transferred by convection to the oil in both vertical and horizontal ducts. Without the oil flow guide washer, the oil flowing in the horizontal duct is not guaranteed.

Oil flows through different passages with different rates [7], and winding temperature doesn't vary strictly linearly with its height. But for the sake of simplifying the calculation, the linear relation is assumed; it works because the real winding temperature distribution is nearly linear and enough margins are given in practical designs. Variation of temperature with winding height is also close to a linear distribution in the case of forced oil cooling. For the same reason of simplifying the calculation, equal amount loss is assumed to be produced in each cable, even though eddy current loss in each cable is slightly different, due to the eddy current loss being much smaller compared to I^2R loss. The extra winding gradient caused by extra loss generated by radial leakage flux at a critical location, such as the top of winding, is considered by hot spot factor.

The winding gradient, $\Delta\Theta_{wdg\,grd}$, consists of two components. One is temperature drop in insulation paper of winding cable $\Delta\Theta_p$. The other is temperature drop from the paper surface to surrounding oil $\Delta\Theta_s$, as shown in Figure 7.9.

$$\Delta\Theta_{wdggrd} = \Delta\Theta_p + \Delta\Theta_s \qquad (7.20)$$

The temperature drop across the insulation paper of the cable caused by conduction is

$$\Delta\Theta_p = q \cdot \left(\delta / \lambda_p\right) \qquad (7.21)$$

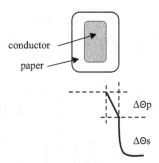

FIGURE 7.9   Temperature drop from winding conductor to surrounding oil.

where $q = W/S$ is heat flux density per unit transfer surface, W/m², W is the loss produced by the winding, S is the cooling area corrected by spacer covering factor, δ is thickness of one side paper insulation, m and λ_p is the thermal conductivity of insulation paper which is $\lambda_p = 0.0465\Theta_p^{0.279}$. To simplify the calculations, the thermal conductivity is chosen to be 0.1395 W/m °C at 55°C, 0.15 W/m °C at 75°C. The temperature drop from the outer surface of insulation paper to oil is by convection, which is

$$\Delta\Theta_s = \frac{q}{\alpha} \qquad (7.22)$$

where q is the same as in Equation (7.21), and $\alpha = A \cdot q^n$ is the surface heat transfer coefficient which depends on heat flux density q, W/m² °C, where A and n are constants. Empirical formulas are achieved based on test results. With the oil flow guide washer, the winding gradient is

$$\Delta\Theta_{wdg\ grd} \approx \frac{\delta \cdot q}{0.15} + \frac{q^{A1}}{B1} \qquad (7.23)$$

where $A1$ and $B1$ are constants. At ONAN rating, for most of units the differences between test and calculated values are in the range of −4~+4°C. At ONFA rating the differences are in the range of −4~+6°C. Also, at the ONFA stage, other factors, such as winding radial build, radial spacer thickness and zigzag oil flow passage height, when not chosen properly, contribute 2~3°C to the gradient. The production defects also cause higher winding gradients. The average winding rise is

$$\Delta\Theta_{wdg} = \Delta\Theta_{wdg\ grd} + \Delta\Theta_{oil\ in\ wdg} \qquad (7.24)$$

Thermal analysis of a transformer is to calculate the temperatures of oil and windings to ensure the top oil rise $\Delta\Theta_{top\ oil}$ and winding rise $\Delta\Theta_{wdg}$ within limits specified either by customer or by standard. The winding hot spot temperature rise, $\Delta\Theta_{HS,}$ is

$$\Delta\Theta_{HS} = K_{HS} \cdot \Delta\Theta_{wdg\ grd} + \Delta\Theta_{top\ oil} \qquad (7.25)$$

The hot spot temperature is one of the key factors determining transformer service life. The temperature rises of lead cables, bushings and switches should be lower than the winding rise which they connect to, to ensure that only the winding is the determining factor on unit service life. The temperature rises of core surfaces, tie-plate

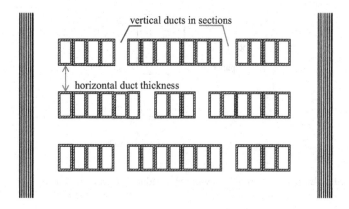

FIGURE 7.10 An example of cooling ducts in winding radial build.

and frames of core, and tank wall should also be checked to be below certain temperatures to avoid oil gassing.

In larger transformers, besides ducts under and over winding sections, sometimes vertical oil ducts are inserted into the sections, as shown in Figure 7.10, to improve the cooling condition of the winding. All of these ducts must be placed asymmetrically with each other in each section, in order to make oil flow in the horizontal ducts. Since there is no oil flow guide washer, the horizontal duct should be wide enough to make oil flowing happen. Generally speaking, the horizontal duct thickness shall be minimum 8% of the section width.

7.4.2.2 Winding Gradient with Directed Forced Oil Flow

With directed forced oil flow, the surface heat transfer is improved while the temperature drop in the insulation paper is unchanged, as shown in Figure 7.11. The surface temperature drop decreases to such an extent at oil velocity of a few dm/second that it is not worth it to increase the oil velocity further, because this would bring an unnecessary increase of pumping working load with less further decrease of the surface temperature.

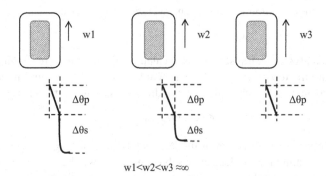

FIGURE 7.11 Winding gradient with different oil flow velocities of w1, w2 and w3.

REFERENCES

1. IEEE Std C57. 91-2011, *IEEE Guide for Loading Mineral-Oil-Immersed Transformers and Step-Voltage Regulators*, New York, USA, 7 March 2012.
2. IEEE std C57.12.00™-2006, *IEEE Standard for General Requirements for Liquid-Immersed Distribution, Power, and Regulating Transformers*, New York, USA, 15 September 2006.
3. K. Karsai, et al., *Larger Power Transformers*, Elsevier Science Publishing Company Inc., Amsterdam, Oxford, New York, Tokyo, 1987.
4. IEC 60076-2, *Loading Guide for Oil-Immersed Power Transformer*, Geneva, Switzerland, 1993.
5. IEEE Std C57.119-2001™, *IEEE Recommended Practice for Performing Temperature Rise Tests on Oil-Immersed Power Transformers at Loads Beyond Nameplate Ratings*, New York, USA, 12 March, 2002.
6. Hasse Nordman, *Fiber Optics for Core Type Transformers, Tutorial Session IEEE Standards Meeting*, Munich, 21 March 2013.
7. Y. Nakamura, et al., Incompressible flow through multiple passages, *Numerical Heat Transfer, Pt. A*, Vol. 16, Issue 4, Dec. 1989, p. 451–465.

8 Short-Circuit Obligation

A transformer is subject to five types of current during its service life:

- Transient in-rush current in primary winding when it is switched in.
- Steady no-load current in primary winding when it has no load on the secondary side.
- Steady load currents in all windings when a load is connected to the secondary winding.
- Transient short-circuit currents in all windings during external short-circuit event.
- Geomagnetically induced currents, which are possible when the winding has a grounding point.

Only the transient short-circuit currents, consequential forces and request withstand strength are discussed in this chapter. A transformer may go through an external short-circuit event several times during its service life. During these events, winding conductors get much higher forces than that under normal service conditions. These forces may deform the windings if their mechanical structure is not robust when it is poorly designed or manufactured. With service years accumulating, the winding mechanical structure is not as tight as the original one, and conductor insulation papers have aged and become brittle. Short-circuit accidents not only probably deform the windings easily but also could tear off the paper insulations, resulting in dielectric breakdown. The force acting on a conductor of length dl carrying current I, is $\vec{df}$, which is governed by the Lorentz force law

$$\vec{df_f} = \overrightarrow{Idl} \times \vec{B} = S \cdot dl \cdot \left(\vec{J} \times \vec{B} \right); \vec{f_f} = \left(\vec{J} \times \vec{B} \right) \tag{8.1}$$

where f_f is the force on unit volume, S is the conductor cross-sectional area, J is current density and B is flux density. The force f is perpendicular to the plan made by J and B, and its direction follows the "left-hand rule". As can be seen from Equation (8.1), the force f on unit volume or stress on winding conductor depends only on the current density in conductor and flux density where the conductor is located. Figure 8.1 shows a typical leakage flux pattern of two-winding units. In the largest part of the windings, the leakage fluxes have only an axial component; as a result, only radial forces exert on the winding conductors. Since the real direction of current in each winding is opposite to another, the radial force directions of both windings are opposite each other, too; that is, the inner winding has an inward radial force, the outer winding an outward radial force. At both ends of the windings, the flux lines bend because both axial and radial components exist in these areas. Besides the radial forces, axial

167

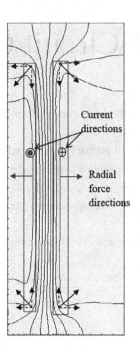

FIGURE 8.1 Leakage flux pattern of a two-winding unit.

forces generated by radial flux components exert on the winding ends; the force direction is discussed later. The rms and peak values of leakage flux density, B and B_{max}, in the gap between windings, are

$$B = \frac{\mu_0 \cdot NI}{H_{wdg}}; \; B_{max} = \sqrt{2}\,\frac{\mu_0 \cdot NI}{H_{wdg}} \tag{8.2}$$

where $\mu_0 = 4\pi \times 10^{-7}$H/m, NI is product of number of turns and rated current of one winding, H_{wdg} is average winding magnetic height. It is seen from Equations (8.1) and (8.2) that the force is proportional to the square of winding current. During short-circuit events, the windings and their leads usually carry about 10~20 times of the rated load current, meaning that the mechanical forces under a short-circuit situation is about 100~400 times of forces under normal operation. Understanding the effects of external short-circuit failures on windings is essential to designing and building windings and their leads mechanically strong enough to survive through the short-circuit events without deformity or damage.

8.1 SHORT-CIRCUIT EVENTS

Under normal operation, the instant voltage, $v(t)$, and current, $i(t)$, of a transformer winding are

$$v(t) = \sqrt{2}\,Vsin(\omega t + \varphi); \; i(t) = \sqrt{2}\,Isin(\omega t + \varphi - \phi) \tag{8.3}$$

where V and I are effective values of the voltage and the current respectively, $\omega = 2\pi f$, is angular frequency where f is frequency; φ is angle of the voltage $v(t)$ at $t = 0$, the moment that short-circuit occurs. $\phi = \arctan\left(\omega L/R\right)$ is the phase angle between the voltage and the current, where L is reactance of the system and transformer, R is resistance of the system and transformer. In a power system, the reactance is much larger than the resistance; it makes $\phi \approx \pi/2$. After the short-circuit event occurs, the current consists of two components, as shown in Figure 8.2; one is sinusoidal with time, another is exponential decay with time.

$$i(t) = \sqrt{2}\, I_{SC}\left[\sin\left(\omega t + \varphi - \phi\right) - e^{-\frac{t}{\tau}}\sin\left(\varphi - \phi\right)\right]; \quad I_{SC} = \frac{100}{\left(Z + Z_s\right)}\cdot I \qquad (8.4)$$

where I_{SC} is effective short-circuit current in winding, I is winding rated effective current and Z and Z_s are transformer and system impedances in percentage respectively, $\tau = L/R$. From Equation (8.4) it can be seen that when $\varphi = 0$ or $\varphi = \pi$, when short-circuit happens at zero instant value of the voltage, the short-circuit current has its maximum value as

$$i_m(t) \approx \sqrt{2}\, I_{SC}\left[\sin\left(\omega t - \frac{\pi}{2}\right) + e^{-\frac{t}{\tau}}\right] \qquad (8.5)$$

Under such a situation, when $t = 1/(2f)$, the short-circuit current reaches its peak value

$$I_{SC\,peak} \approx \sqrt{2}\, I_{SC}\left[1 + e^{-\frac{\pi R}{X}}\right] = K I_{SC} = \sqrt{2}\cdot k \cdot r \cdot I \qquad (8.6)$$

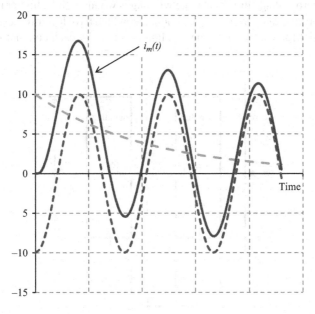

FIGURE 8.2   Short-circuit current.

where
$$K = \sqrt{2}\left[1 + e^{-\frac{\pi R}{X}}\right] = \sqrt{2}\,k; \quad r = \frac{100}{\left(Z + Z_s\right)} \tag{8.7}$$

k is usually called asymmetrical factor, K is called peak factor. A transformer must be designed and built to withstand the forces generated by i_{peak}. The faults in a remote station supplied through a transformer are damped by the secondary line impedance to harmless amplitude when the distance is only a few kilometers. At the opposite extreme, a unit for auxiliary power supply in a generating station receives all that faults by its own impedance.

8.2 RADIAL AND AXIAL ELECTROMAGNETIC FORCES

The study of short-circuit forces usually focuses on radial component and axial component separately instead of on the total force. Radial force is generated by axial component of leakage flux, while axial force is generated by radial component of leakage flux.

8.2.1 RADIAL FORCE

The axial leakage flux density distribution in the middle of winding assembly, where ampere-turn balances well between two windings, is shown in Figure 8.3. The conductors adjacent to the main leakage flux channel, that is the gap between the windings, have the highest radial forces because the leakage fluxes are highest in these locations; the conductors on the other side of the windings have zero radial forces because of zero leakage flux. When the winding is wound tightly, higher radial forces on conductors are transferred next to conductors which have lower forces. As a result, the total radial force exerted on the winding is shared by each conductor evenly, so

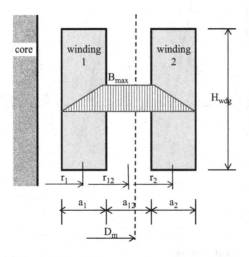

FIGURE 8.3 Leakage flux in two windings.

the average radial force on one electrical turn (can be several conductors) per unit circumference length based on Equation (8.1) is

$$F_{unit\ length} = \left(B_{max} \cdot I_{peak}\right)/2 \tag{8.8}$$

where I_{peak} is peak amplitude of the winding rated current I. The total radial force on a complete winding, if ignoring the reduction of the axial flux at winding ends, is

$$F_{rad} = F_{unit\ length} \cdot \pi \cdot D_{mwdg} \cdot N = \frac{\mu_0}{2} \cdot \frac{\sqrt{2}\left(NI\right)}{H_{wdg}} \cdot I_{peak} \cdot \pi \cdot D_{mwdg} \cdot N$$

$$F_{rad} = \frac{\mu_0}{2} \cdot \frac{\left(NI_{peak}\right)^2}{H_{wdg}} \cdot \pi \cdot D_{mwdg} \tag{8.9}$$

where D_{mwdg} is mean diameter of the winding, m. F_{rad} is the maximum radial force generated by current I_{peak}, Newton. Under short-circuit situation, the maximum short-circuit current is $I_{SC,peak} = \sqrt{2} \cdot k \cdot r \cdot I$, then maximum radial force under short-circuit, $F_{sc\ rad}$, is

$$F_{sc\ rad} = \mu_0 \cdot \frac{\left(NI\right)^2}{H_{wdg}} \cdot \pi \cdot D_{mwdg} \cdot \left(r \cdot k\right)^2 \tag{8.10}$$

The resultant tensile and compressive tangential stress, $\sigma_{sc.rad}$, is

$$\sigma_{sc\ rad} = \frac{P \cdot D_{mwdg}}{2 \cdot RB} \tag{8.11}$$

The related geometric dimensions of winding are shown in Figure 8.4. P is pressure in the winding surface.

$$P = \frac{F_{sc\ rad}}{\pi \cdot D_{mwdg} \cdot H_{wdg}} \tag{8.12}$$

P is in N/m², so

$$\sigma_{sc\ rad} = \frac{\mu_0}{2} \cdot \left(\frac{NI}{H_{wdg}}\right)^2 \cdot \left(r \cdot k\right)^2 \cdot \frac{D_{mwdg}}{RB} \tag{8.13}$$

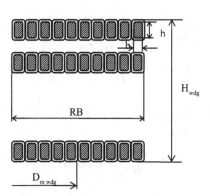

FIGURE 8.4 Winding conductor arrangement.

Here, the number of turns, N, can be expressed by number of section, N_{sec}, times the number of turns per section, $N_{turn/sec}$.

$$N = N_{sec} \cdot N_{turn/sec} \qquad (8.14)$$

The winding radial build of bare conductor, RB, is

$$RB = b \cdot N_{turn/sec} = \frac{b \cdot N_{turn/sec} \cdot H_{wdg} \cdot J}{H_{wdg} \cdot J}$$

$$RB = \frac{b \cdot N_{turn/sec} \cdot N_{sec} \cdot h \cdot J}{H_{wdg} \cdot J \cdot F_{s\,axial}} = \frac{NI}{H_{wdg} \cdot J \cdot F_{s\,axial}} \qquad (8.15)$$

where J is winding current density, A/m^2; $F_{s\,axial} = (N_{sec} \times h)/H_{wdg}$, is winding axial space factor, the ratio of winding total bare conductor height to winding overall height. The winding overall height is the sum of total bare conductor height, total conductor insulation paper thickness and total radial spacer thickness. Usual values of $F_{s\,axial}$ are in the range of 0.4~0.6. Substitute Equation (8.15) into Equation (8.13)

$$\sigma_{sc\,rad} = \frac{\mu_o}{2} \cdot \frac{NI}{H_{wdg}} \cdot J \cdot D_{mwdg} \cdot F_{s\,axial} \cdot (r \cdot k)^2$$

$$\sigma_{sc\,rad} = \frac{\mu_0}{2\pi} \cdot \frac{N \cdot \pi \cdot D_{mwdg} \cdot (b \cdot h)}{H_{wdg}} \cdot J^2 \cdot F_{s\,axial} \cdot (r \cdot k)^2$$

$$\sigma_{sc\,rad} = \frac{\mu_0}{2\pi} \cdot \frac{V}{H_{wdg}} \cdot J^2 \cdot F_{s\,axial} \cdot (r \cdot k)^2$$

$$\sigma_{sc\,rad} = \frac{\mu_0}{2\pi\delta} \cdot \frac{W}{H_{wdg}} \cdot J^2 \cdot F_{s\,axial} \cdot (r \cdot k)^2 \qquad (8.16)$$

where $V = N \cdot (\pi \cdot D_{mwdg}) \cdot (b \cdot h)$, is winding copper volume per leg, m^3; $\delta = 8900$ kg/m^3, copper mass density; W is winding copper weight per leg, kg. The I^2R loss of a copper winding at 75°C is $P_{DC@75} = 2.36 \times J^2 \times W$, Watts. Equation (8.16) then can be expressed as

$$\sigma_{sc.rad} = \frac{4\pi \times 10^{-7}}{4\pi \times 8.9 \times 10^3 \times 2.36} \cdot \frac{P_{DC@75}}{H_{wdg}} \cdot F_{s\,axial} \cdot (r)^2 \left(\sqrt{2} \cdot k\right)^2$$

$$\sigma_{sc.rad} = 4.76 \times 10^{-12} \times \frac{P_{DC@75}}{H_{wdg}} \cdot F_{s\,axial} \cdot (r)^2 \left(\sqrt{2} \cdot k\right)^2 \qquad (8.17)$$

where $\sigma_{sc.rad}$ is in N/m^2. When $P_{DC@75}$ is of kW, H_{wdg} is of mm as usually used in design, Equation ((8.17) then is

$$\sigma_{sc.rad} = 4.76 \times \frac{P_{DC@75}}{H_{wdg}} \cdot F_{s\,axial} \cdot (r)^2 \left(\sqrt{2} \cdot k\right)^2 \qquad (8.18)$$

Here, $\sigma_{sc.rad}$ is in N/mm². From Equation (8.18), the radial stress of one winding is related to its I^2R loss and winding height. The inner winding is under inward force which tends to crush the winding towards the core. The phenomena occurring under inward compressive forces will be discussed next. The outer winding is under outward force which tends to burst the winding. When proof stress of copper is adequate, the outward burst force doesn't present a series problem normally, except that the outer winding has a small number of turns per section, which can be the case when the outer winding is a helical tap winding, another LV helical winding or a disc winding having a smaller number of turns. In such situation, it is important to take necessary measures to resist the burst force.

8.2.2 Axial Force

Two types of axial forces exist. One is produced by the radial flux component at winding ends which, as shown in Figure 8.1, produces axial compressive forces in the winding. The directions of these forces are toward the center of the winding. These forces exist even there is no gap in windings and both windings have the same height, which makes a well-balanced mmf (magneto motive force, or ampere-turn distribution). For a two-winding unit, the magnetic energy stored in the winding system is

$$W = \frac{\pi \cdot \mu_0 (NI)^2}{H_{wdg}} \left(\frac{1}{3} r_1 a_1 + r_{12} a_{12} + \frac{1}{3} r_2 a_2 \right) \tag{8.19}$$

where the dimensions used in the equation are shown in Figure 8.3. The axial force exerting on winding pair per leg, F_{ax}, is

$$F_{ax} = -\frac{\partial W}{\partial H_{wdg}} = \frac{\pi \cdot \mu_0 \cdot (NI)^2}{2H_{wdg}^2} \left(\frac{D_1 a_1}{3} + D_{12} a_{12} + \frac{D_2 a_2}{3} \right) \tag{8.20}$$

where $D_1 = 2r_1$, $D_2 = 2r_2$, $D_{12} = 2r_{12}$, the mean diameters of winding 1, winding 2 and the gap in between respectively. This is axial force under rated current, I. Under the short-circuit situation, this force becomes

$$F_{sc\cdot ax} = \frac{\mu_0}{2} \cdot \left(\frac{NI}{H_{wdg}} \right)^2 \cdot \pi \cdot \left(D_{12} \cdot a_{12} + \frac{D_1 \cdot a_1 + D_2 \cdot a_2}{3} \right) \cdot (2 \cdot K_g - 1) \cdot r^2 \cdot \left(\sqrt{2} \cdot k \right)^2 \tag{8.21}$$

where $F_{sc.ax}$ is total axial force per leg, Newton; H_{wdg} is average axial height of windings, mm; D_1, D_2, D_{12}, a_{12}, a_1 and a_2 are in mm; and K_{rog} is Rogowski factor.

$$K_g \approx 1 - \frac{a_1 + a_{12} + a_2}{\pi \cdot H_{wdg}} \tag{8.22}$$

As can been seen from Equations (8.10) and (8.21), a large rated power unit has high axial and radial forces compared to a small rated power unit provided that both units have the same impedance. With a unit having three ratings, such as ONAN, ONAN/ONAF and ONAN/ONAF/ONAF, the short-circuit forces of each rating are very close to each other because the impedance increases with rated power. Both radial and axial forces are inversely proportional to the winding axial height. When, for example, transportation limits or size of unit make it necessary to reduce winding height, higher short-circuit forces are then expected.

Figure 8.5 shows axial forces at conductors of inner winding and outer winding of a two-winding unit, in case (a) the ampere-turn is well balanced because at maximum tap position, all turns including tap turns of HV winding are in circuit.

Another type of axial force exists due to unbalanced mmf (magneto motive force, or ampere-turn distribution) between windings, which is caused by different winding heights and gaps in winding. These axial forces generated by such radial flux tend to separate the inner winding and outer winding in an axial direction; if the axial force pulls the inner winding up, the force at the outer winding pushes it down. As the result, there are forces exerting on top and bottom clamping rings, sometimes called end thrust forces. For a two-winding unit, if the inner winding such as LV winding has no gap, while the outer winding such as HV winding has a tap gap in the middle, at minimum voltage tap, all tap sections and turns do not carry current, the gap has its largest size as shown in Figure 8.5b, the radial leakage flux in such gap generates axial force. To reduce such forces, LV winding turns in the area facing the tap area of HV winding should be reduced to try to balance the ampere-turn between the two windings. Two methods can be used to achieve this goal. One method is to make the so-called compensation gap in LV winding, facing HV tap gap. The other method is to make this LV winding area consist of few turns separated by thicker spacers.

Yoke laminations attract the leakage fluxes from winding ends, make them more axial directed; this could reduce axial forces at the winding ends. However, only a portion of winding is above the bottom yoke and below the top yoke directly; the rest is outside that window, and this portion of winding has higher axial forces. Design should be verified to withstand these higher axial forces. In ideally symmetrical windings (with completely balanced ampere-turns), the axial forces tend to compress the windings and are in equilibrium with each other; no force is exerted on structural parts outside the windings. However, difference in winding heights always exists because of winding materials and production, the HV winding tap gap size changes with different tapping positions which means one size of LV compensation gap cannot make perfect balances for all HV taps, and such unbalanced ampere-turns generate axial forces tending to increase the existing asymmetry. Consequently, there are forces exerting on the structural components of the transformer, such as clamping rings and clamping structure. The transformer structure reaction to axial force is quite different compared to radial force, due to the structure being less rigid or more elastic in the axial direction than in the radial direction.

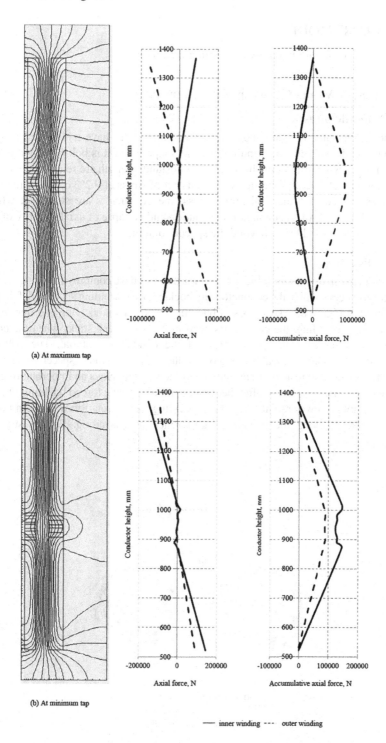

(a) At maximum tap

(b) At minimum tap

—— inner winding --- outer winding

FIGURE 8.5 Axial forces on a two-winding unit.

8.3 FAILURE MODES

Several types of failure in reality are summarized here.

8.3.1 FAILURE MODES CAUSED BY RADIAL FORCES

8.3.1.1 Tensile Stress

Conductors in outer winding subjected to outward forces have tensile stress. This tensile stress doesn't change winding circular shape when it is below the withstand strength; the design criteria of short-circuit withstand strength are related only to the conductor's mechanical properties such as its proof stress at 0.2% offset. If the tensile stress exceeds the material proof stress, the conductors are stretched, which could result in the broken conductor's insulation and collapse of axial stability of the winding due to a local bulge beyond the spacer contour.

8.3.1.2 Buckling

Conductors in inner winding subjected to inward forces have compressive stress. When such stress exceeds limits, the conductors are buckled or the winding shape is deformed, as sketched in Figure 8.6. Two types of buckling exist: forced buckling and free buckling. In forced buckling, the conductors bend inwards between sticks when the compressive stress exceeds the proof stress of the conductor material. The collapse mechanism is the same as axial bending of conductor between radial spacers. The number of sticks placed at insider circumference of the winding plays an important role.

Free buckling occurs more often because, compared to the forced buckling, it happens under much lower radial force. Free buckling has no relation with the number of sticks placed along the insider circumference of a winding. Factors such as

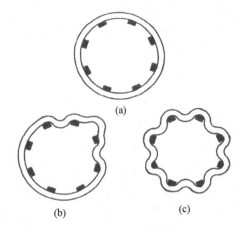

(a) original winding shape
(b) shape after free buckling
(c) shape after forced buckling

FIGURE 8.6   Inner winding deformation by buckling.

tightness, initial eccentricity and diameter of winding, and conductor geometry are decisive factors on free buckling critical stress. The dimension and the proof stress of the conductor, the bonding effect of CTC or multi-wire conductors contribute greatly to resistance to free buckling deformation. Design concept regarding the free buckling shall be that the windings are designed to be completely self-supporting. It sometimes needs to reduce current density in winding conductors to achieve self-support; as a consequence, more copper is needed.

8.3.1.3 Spiraling

Another type of failure mode caused by the radial (as well as axial) compressive forces is known as "spiraling", sketched in Figure 8.7. It is a deformation pattern that typically happens in the helical type of winding. When helical winding has high pitch, such as one electrical turn consisting of several cables axially stacked, the risk of spiraling is high. When LV winding consists of two physical helical windings, such as LV winding of large GSU, the spiraling is confined to the winding adjacent to the main leakage flux channel, which has higher radial forces compared to the inner winding. If the winding is made by epoxy-bonded CTCs, the stress limit to avoid the spiraling may be the key factor in comparison with the critical stress for buckling.

8.3.2 FAILURE MODES CAUSED BY AXIAL FORCES

8.3.2.1 Tilting

Under high axial forces, the conductors lost axial stability and tilt, as shown in Figure 8.8. Generally, winding having a thin conductor, large diameter and fewer strands in a radial build are exposed to a higher risk of tilting. In addition, the coverage of the radial spacer plays a role since greater coverage provides high resistance to tilt. A conductor with a large corner radius inclines to tilt more easily than one with a small corner radius. A high conductor hardness degree improves critical tilt strength, a concept that will be discussed in the following section. All discussion to this point is applied to a strand conductor such as single, twin or triple wires and non-epoxy-bonded CTC. For winding made of epoxy-bonded CTC, since all strands are bonded

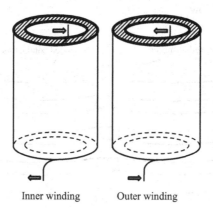

Inner winding   Outer winding

FIGURE 8.7 Spiraling failure.

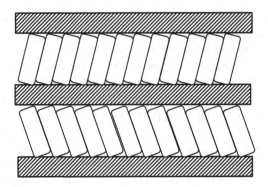

FIGURE 8.8 Conductors between spacers tilt.

together making CTC like a solid bar, there is no design constraint regarding the tilt. In fact, epoxy-bonded CTC are extremely resistant to tilting regardless of the conductor hardness degree.

The pressure applied to winding made by epoxy CTC in oven needs to be checked. If the pressure is too high, the strands could tilt before the epoxy cured, making the cable permanently deformed.

8.3.2.2 Axial Bending
Under high axial forces, the conductors are bended between spacers, as shown in Figure 8.9. The bending stress will be discussed in the following section.

8.3.2.3 Telescoping
Telescoping is an axial collapse of winding. Telescoping drives some turns axially past adjacent turns because the radial looseness caused by stretched conductors.

8.3.2.4 Collapse of Winding End Supports
The end insulation structures between yoke and winding could be broken if the end thrust stress is higher than the strength of the material, or the stability of the end insulation structure is low.

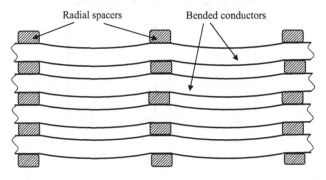

FIGURE 8.9 Axial bending between radial spacers.

Under axial forces, winding conductors could move (vibrate) if the clamping pressure on the winding set is not high enough to hold the conductors steady, or if the end insulation structure is soft. A mechanically strong end insulation support structure and a robust clamping structure are important to prevent this type of vibration. The robust clamping structure is capable of both absorbing the axial forces and maintaining long-term effective pressure on the winding set. With accumulation of service years, winding insulations such as conductor paper and spacer age and become soft and brittle; this results in reduction of the initial pressure applied on the winding set and making the windings loose. In such cases, vibrations caused by short-circuit forces could easily rip off the paper, bringing in electrical breakdown after a short-circuit event and eventual failure. To keep the winding set under a certain pressure after years of service, one option is that a high initial clamping pressure is applied. It is thought that although the clamping pressure decreases with time, the remains of the initial clamping pressure are still enough to hold the windings tight. Also, high clamping pressure increases the natural mechanical resonance frequencies of the winding itself. Other types of failures happening to winding lead, crossover and transposition will not be discussed here in detail.

The following points may be considered for basic robust design:

- Reduce short-circuit forces by adjusting winding heights, diameters and current density in the conductor. From the previous discussion, the taller the winding, the lower the forces. However, the winding height is limited by other factors such as the impedance and shipping height of unit. Reduction of the current density can reduce the short-circuit forces, but more material, bigger overall sizes and more labor are added to the cost.
- Increase the proof stress of conductor. Higher proof stress makes the conductor stiffer against deformation in short-circuit events. Keep in mind that the stiffer the conductor is, the more difficult it is to be wound, the winding quality may be sacrificed.
- Using epoxy-bonded CTC. The epoxy provides CTC with rigidity almost comparable to a massive bar which has an equivalent cross-sectional area. It improves both radial and axial strengths, and it is the reason epoxy-bonded CTC are recommended and widely used in large units. The epoxy's bonding strength plays a critical role in such improvement. Regarding the change in epoxy-bonding strength with temperature, test results from some manufactures show that the bonding strength at 125°C is about 50% of the strength at 25°C, it is also found from experiments that there is no significant decay in the bonding strength from 100°C to 120°C provided that the epoxy curing process has been performed properly [1,2]; 50% bonding strength at 120°C is usually accepted in the industry.
- Increasing impedances which are related to tertiary winding. With a large unit, especially an autotransformer, the tertiary winding is sometimes the weak point for both three-phase fault and single-phase fault. Increasing impedances between HV to tertiary, LV to tertiary can reduce short-circuit currents and the related forces in all windings. Two ways exist to increase the impedances. One is to increase the distances between tertiary winding and other windings; this

method is simple but makes whole core and coil assembly larger. The other method is to use the current limit reactor, which is connected to tertiary winding, to increase the impedance. The current limit reactor could sometimes be less expensive than the first method, it is worth it to compare two methods on each actual design and select the better one to fit.

8.4 SHORT-CIRCUIT FORCES IN SPECIAL TRANSFORMERS

For an autotransformer, the neutral is solid grounded in most cases, and the grounded neutral makes single phase-to-ground fault possible. The impedances of autotransformers are usually smaller than two-winding units. These two factors result in greater short-circuit forces in autotransformer windings. The study on short-circuit forces on design stage should be conducted thoroughly.

For large GSU (generator step up) transformers, shipping height is often the factor limiting core and coil assembly height, although the core height can be reduced with reduced yoke height by adopting five-leg core, it still cannot sometimes meet the height limit, so reducing short-circuit forces by increasing winding height is not an option here. Also, when relatively small impedances are required for large units, this cannot be achieved by increasing the winding height for the same reason mentioned. To solve this issue, HV winding is split into two parts which are placed on either side of LV winding (HLH). In fact, this arrangement is very effective in reducing impedance value. This winding arrangement is sometimes called radial split. The transformers for auxiliary power supply in power stations sometimes have two sets of secondary winding with one primary winding; it is another type of radial split, as shown in Figure 8.10a. LV windings are placed on either side of the HV winding (LHL). Either HV winding or LV winding can be split into two equal parts, or into unequal parts with 30% and 70% as extremes. Regarding the short-circuit forces, the innermost and outmost windings are subjected to compressive and tension forces respectively, while the winding in the middle is subjected to much fewer radial forces. When the outmost winding has a thin radial build due to few turns such as tap winding, its tension stress should be studied. Preventive measures, for example keyed-in sticks around the winding's outside surface should be taken to prevent the winding burst. The middle winding is subjected to higher axial force because of higher radial flux components at its ends.

There is another arrangement for two LV and one HV windings, in which the LV windings are stacked axially one on the top of the other, and the HV winding is also split into two parts which are connected in parallel. It is called axial split, as shown in Figure 8.10b. For axial split winding arrangement, great care has to be taken in short-circuit performance. When one of LV windings is short-circuited, while the other LV is open, there is still a certain amount current flowing through the HV winding part facing the opened LV winding, because the impedances of HV2-LV1 or HV1-LV2 are not infinite, although they are very high. The ampere-turn is not entirely balanced in such a case, and this may cause higher short-circuit forces than usual. The same phenomenon causes extra eddy current loss on metallic parts, or circulating current if both HV halves are connected to

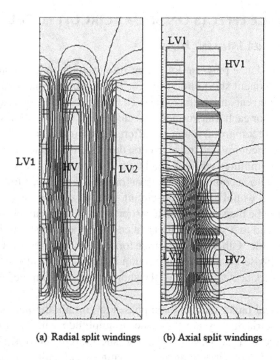

(a) Radial split windings   (b) Axial split windings

FIGURE 8.10 Leakage flux distributions.

one tap changer. To prevent such circulating current flowing in HV winding, two tap changers should be used.

8.5 SHORT-CIRCUIT CURRENT CALCULATION

For a two-winding three-phase unit, the fault of three-phase-to-ground usually gives the highest short-circuit current. Its rms value is

$$I_{SC} = \frac{100}{Z + Z_s} \cdot I; \ Z_s = \frac{100 \cdot S_r}{S} \tag{8.23}$$

where Z is impedance of transformer, %; Z_s is impedance of system, %; I is rated current of transformer, A; and S_r and S are transformer rated power and system's short-circuit apparent power respectively. Keep in mind that the short-circuit force is proportional to the current square and that current is the initial peak asymmetrical current, $I_{SC\,peak}$, shown in Equation (8.6), not rms value, I_{sc}. The relation between these two currents is

$$I_{SC\,peak} = k \cdot \sqrt{2} \cdot I_{SC} \tag{8.24}$$

where k is per Equation (8.7). RMS value is used here for convenience in discussion.

8.6 IMPEDANCE EFFECTS ON SHORT-CIRCUIT FORCE

8.6.1 Transformer Inherent Impedance

Transformer impedance plays an important role in magnitudes of short-circuit current and the consequent short-circuit forces. The higher the impedance is, the lower the short-circuit current and the forces are. There is a so-called "natural" short-circuit impedance value for each transformer. By selecting it, the overall design can be well balanced with best compromise among different requirements.

For small and medium-size transformers, the system to which they are connected can be treated as infinite; whereas for larger transformers, the system impedance should be taken into account. When a transformer is connected to other equipment whose impedance would limit short-circuit current significantly, that equipment impedance should be then considered in a short-circuit force calculation. If the transformer is connected to a generator by isolated-phase bus, no short-circuit fault happens on the LV side. Otherwise, if a transformer is connected to a generator by a non-segregated phase bus, or the connection includes a circuit breaker, or there is other equipment connected to the bus, the short-circuit fault on the LV side has to be considered.

For symmetrical faults such as three phases to ground, the fault currents in each phase are symmetrical; they are the same in amplitude with each other and 120° apart, like balanced three phase loads. The short-circuit current is calculated by Equation (8.23) for two winding units. For asymmetrical faults such as single phase-to-ground, the amplitude of short-circuit current in each phase is different from other phases; they are not 120° apart anymore, they are asymmetrical. In order to calculate the short-circuit current in each phase under the asymmetrical fault, the symmetrical-component analysis method is used. By this method, the asymmetrical currents are decomposed into three symmetrical components, positive sequence, negative sequence and zero sequence components, which are calculated by using positive, negative and zero sequence impedances. Based on these currents, the total short-circuit current in each phase is then composited. For any static-type equipment, such as transformers, $Z_+ = Z_-$, Z_+ is positive sequence impedance and Z_- is negative sequence impedance. Many books and standards give the calculations of asymmetrical fault currents [1,3,4,5], Table 8.1 provides the summary; the detailed deductions of the calculation formula are omitted here.

For a three-winding (three-circuit) unit, the impedances can be represented by assuming that each winding has its own impedance, although in reality the impedances exist between pairs of windings. The reason for such mathematical treatment is only to make the calculation simple. The winding impedances are calculated as

$$Z_H = \frac{\left(Z_{HX} + Z_{HY} - Z_{XY}\right)}{2}; \ Z_X = \frac{\left(Z_{HX} + Z_{XY} - Z_{HY}\right)}{2};$$

$$Z_Y = \frac{\left(Z_{XY} + Z_{HY} - Z_{HX}\right)}{2} \tag{8.25}$$

The calculation of impedances and short-circuit current on each winding shall be based on a common MVA.

TABLE 8.1
Short-circuit Currents

Type of fault	Phase a	Phase b	Phase c
External Neutral Impedance = 0			
Single phase to ground	$I_+ = I_- = I_0 = \dfrac{I}{Z_+ + Z_- + Z_0}; I_a = I_+ + I_- + I_0$	$I_b = 0$	$I_c = 0$
Double-phase (Line-line)	$I_a = 0$	$I_b = \dfrac{-j \cdot \sqrt{3} \cdot I}{Z_+ + Z_-}$	$I_c = \dfrac{j \cdot \sqrt{3} \cdot I}{Z_+ + Z_-}$
Double phase to ground	$I_a = 0$	$I_b = -j \cdot \sqrt{3} \cdot I \cdot \dfrac{Z_0 - a \cdot Z_-}{Z_+ Z_- + Z_+ Z_0 + Z Z_0}$	$I_c = j \cdot \sqrt{3} \cdot I \cdot \dfrac{Z_0 - a^2 \cdot Z_-}{Z_+ Z_- + Z_+ Z_0 + Z Z_0}$
Three phase to ground	$I_+ = \dfrac{I}{Z_+}; I_- = I_0 = 0; I_a = I_+$	$I_+ = a^2 \dfrac{I}{Z_+}; I_- = I_0 = 0; I_b = I_+$	$I_+ = a \cdot \dfrac{I}{Z_+}; I_- = I_0 = 0; I_c = I_+$
External Neutral Impedance = Z_{ex}			
Single phase to ground	$I_+ = I_- = I_0 = \dfrac{I}{Z_+ + Z_- + Z_0 + 3Z_{ex}}$ $I_a = I_+ + I_- + I_0$	$I_b = 0$	$I_c = 0$
Double-phase (Line-line)	$I_a = 0$	$I_b = \dfrac{-j\sqrt{3} \cdot I}{Z_+ + Z_- + Z_{ex}}$	$I_c = \dfrac{j \cdot \sqrt{3} \cdot I}{Z_+ + Z_- + Z_{ex}}$
Double phase to ground	$I_a = 0$	$I_b = -j \cdot \sqrt{3} \cdot I \cdot \dfrac{3Z_{ex} + Z_0 - a \cdot Z_-}{3Z_{ex}(Z_+ + Z_-) + Z_+ Z_- + Z_+ Z_0 + Z Z_0}$	$I_c = j \cdot \sqrt{3} \cdot I \cdot \dfrac{3Z_{ex} + Z_0 - a^2 \cdot Z_-}{3Z_{ex}(Z_+ + Z_-) + Z_+ Z_- + Z_+ Z_0 + Z Z_0}$
Three phase to ground	$I_+ = \dfrac{I}{Z_+ + Z_{ex}}; I_- = I_0 = 0; I_a = I_+$	$I_+ = a^2 \dfrac{I}{Z_+ + Z_{ex}}; I_- = I_0 = 0; I_b = I_+$	$I_+ = a \cdot \dfrac{I}{Z_+ + Z_{ex}}; I_- = I_0 = 0; I_c = I_+$

Where I is rated winding current, I_a, I_b and I_c are short-circuit current rms values of phase a, b and c respectively, Z_+, Z_- and Z_0 are positive, negative and zero sequence impedances in percentage

$$a = e^{j120} = -\frac{1}{2} + j\frac{\sqrt{3}}{2}; \quad a^2 = e^{j240} = -\frac{1}{2} - j\frac{\sqrt{3}}{2}$$

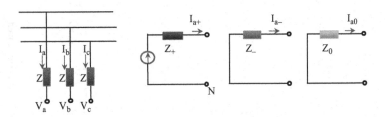

(a) Three phase connections   (b) Sequence network of a phase viewed from fault point

FIGURE 8.11   Sequence networks of a three-phase transformer.

Here, the positive direction of current flow in each sequence network is assumed to be outwards at the faulted or unbalanced point, so the sequence currents are assumed to flow in the same direction, as shown in Figure 8.11. This assumption of current direction must be carefully followed to avoid error when using the equations, even though some of the currents are negative. The procedure of short-circuit current calculation usually is

1. Determine the values of positive, negative and zero sequence impedances of transformer and related systems, as seen from the fault point.
2. Set up sequence networks as viewed from the fault.
3. Make a proper connection between the networks to present the real short-circuit condition.
4. Calculate the sequence currents I_+, I_-, I_0 and total current in each phase.

Example 8.1

A three-phase autotransformer has rated power of 180 MVA with tertiary windings of 37.5 MVA. The impedances discussed here are in percentages. The positive or negative impedances are:

$$Z_{HV\text{-}LV} = 5.47\% \text{ at } 180 \text{ MVA}$$
$$Z_{HV\text{-}TV} = 7.108\% \text{ at } 37.5 \text{ MVA}; 34.12\% \text{ at } 180 \text{ MVA}$$
$$Z_{LV\text{-}TV} = 11.335\% \text{ at } 37.5 \text{ MVA}; 54.41\% \text{ at } 180 \text{ MVA}$$

The system short-circuit apparent powers are

HV system = 50,200 MVA; LV system = 25,100 MVA; TV system = 0 MVA

The ratios of zero sequence impedance to positive sequence impedances are

The unit = 0.85; HV system = 1.0; LV system = 1.0

The positive and negative sequence impedances of each winding per Equation (8.25) are

$$Z_{H+} = Z_{H-} = (5.47 + 34.12 - 54.41)/2 = -7.41\%$$
$$Z_{X+} = Z_{X-} = (5.47 + 54.41 - 34.12)/2 = 12.88\%$$
$$Z_{Y+} = Z_{Y-} = (34.12 + 54.41 - 5.47)/2 = 41.53\%$$

The system impedances are

$$Z_{SH+} = Z_{SH-} = \frac{180}{50,200} = 0.36\%; Z_{SX+} = Z_{SX-} = \frac{180}{25,100} = 0.72\%; Z_{SY} \rightarrow infinite$$

The zero sequence impedances in zero sequence network are

$$Z_{H0} = -7.41 \times 0.85 = -6.30\%; Z_{X0} = 12.88 \times 0.85 = 10.95\%$$
$$Z_{Y0} = 41.53 \times 0.85 = 35.3\%; Z_{SH0} = 0.36 \times 1.0 = 0.36\%;$$
$$Z_{SX0} = 0.72 \times 1.0 = 0.72\%; Z_{SY0} = 0$$

When a single phase-to-ground fault occurs at one of LV phases, the connection of each sequence network is shown in Figure 8.12. The total positive, negative and zero sequence impedances are

$$Z_+ = Z_- = \frac{(12.88 - 7.41 + 0.36) \times 0.72}{12.88 - 7.41 + 0.36 + 0.72} = 0.64\%; Z_0 = \frac{(10.95 - 7.14) \times 0.72}{10.95 - 7.14 + 0.72} = 0.6\%$$

Where $-7.14 = \dfrac{(-6.3 + 0.36) \times 35.3}{-6.3 + 0.36 + 35.3}$. There is no external neutral impedance.

Based on Table 8.1, in the case of single phase-to-ground fault, the short-circuit current of each sequence at fault point is

$$I_+ = I_- = I_0 = \frac{100 \cdot I}{0.64 + 0.64 + 0.6} = 53.19 \cdot I$$

The total fault current is

$$I_a = I_+ + I_- + I_0 = 53.19 \times 3 \times I = 159.6 \times 903.6 = 144214 (A)$$

where I = 903.6 A is LV terminal rated current. The short-circuit currents in each winding are shown in Figure 8.13.

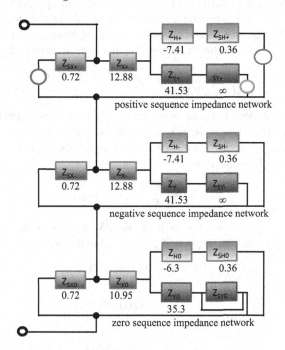

positive sequence impedance network

negative sequence impedance network

zero sequence impedance network

FIGURE 8.12 Impedance network of single-phase fault on LV line.

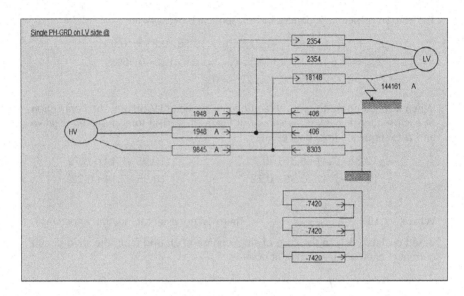

FIGURE 8.13   Short-circuit current in each coil when single LV terminal is grounded.

8.6.2 EXTERNAL NEUTRAL IMPEDANCE IN ZERO SEQUENCE NETWORK

Impedance in neutral does not appear in either positive or negative sequence networks, since three phase currents of either sequence add up to zero at the neutral. However, an equivalent impedance equal to three times the neutral impedance appears in a zero sequence network, since the zero sequence current adds up to three times. For two or three winding transformers, this equivalent impedance is

$$Z_\Omega = \frac{Z_\% \times V_{kV}^2}{3 \times 100 \times MVA}; \qquad Z_\% = \frac{3 \times 100 \times MVA}{V_{kV}^2} \times Z_\Omega \qquad (8.26)$$

where Z_Ω is ohmic value in Ω, $Z_\%$ is percentage value of zero sequence impedance, %, V_{kV} is line-to-line voltage the side of which the impedance is seen, kV, MVA is power rating of three phases, MVA. When LV neutral impedance, $Z_{\Omega LVside}$, is seen, for example, from the HV side, it is

$$Z_{\Omega HVside} = \left(\frac{N_{HV}}{N_{LV}}\right)^2 \cdot Z_{\Omega LVside} \qquad (8.27)$$

Where N_{HV} and N_{LV} are number of turns of HV and LV windings respectively. Autotransformer neutral impedance, Z_Ω, is seen from the primary side as

$$Z_{HO}\% = -\left(\frac{r}{1-r}\right) \cdot \frac{3 \times 100 \times MVA}{V_{kV}^2} \times Z_\Omega$$

$$Z_{XO}\% = \frac{r}{(1-r)^2} \cdot \frac{3 \times 100 \times MVA}{V_{kV}^2} \times Z_\Omega \qquad (8.28)$$

$$Z_{TO}\% = \left(\frac{1}{1-r}\right) \cdot \frac{3 \times 100 \times MVA}{V_{kV}^2} \times Z_\Omega$$

where r is the co-ratio

$$r = \frac{S_d}{S_r} = \frac{V_{HV} - V_{LV}}{V_{HV}} \tag{8.29}$$

S_d is design power or power transformed through magnetic circuit, S_r is rated power, V_{HV} is high voltage, V_{LV} is low voltage.

Example 8.2

1. A 25MVA unit, HV line-to-line voltage is 215.5 kV wye connection, 1466 turns per winding. LV line-to-line voltage is 28 kV zig-zag connection, 110 turns per winding. The calculated positive and negative sequence impedances are $Z_{HL+}\%$ = $Z_{HL-}\%$ = 8.56%. The impedance between zig and zag windings is $Z_{zz}\%=2.62$, the zero sequence impedance seen from the LV side is then $Z_{X0}\%$ = 2.62/3 = 0.87%. The unit's zero sequence impedance is much less compared to wye connected windings; this is why external neutral impedance is usually required for zigzag windings. The zero sequence impedance seen from the HV side is $Z_{X0}'\%$ = $(215.5/28)^2 \times 0.87\%$ = 51.5%. An external impedance of 1.5 Ω is added to the neutral, this neutral impedance in percentage is

$$Z_{EX}\% = \frac{100 \times 25}{28^2} \times 1.5 = 4.78\%$$

The short-circuit apparent powers are LV system =4,500 MVA, HV system =50,000 MVA. The system impedances are

$$Z_{SX+}\% = Z_{SX-}\% = \frac{25}{4500} = 0.56\%; \quad Z_{SH+}\% = Z_{SH-}\% = \frac{25}{50,000} = 0.05\%$$

The ratios of zero sequence impedance to positive sequence impedance of the HV and the LV systems are two, so the zero sequence impedances of LV and HV system are

$$Z_{SX0}\% = 2 \times 0.56\% = 1.12\%; \quad Z_{SH0}\% = 2 \times 0.05\% = 0.1\%$$

For zigzag connection, the fault current of single phase-to-ground may be higher than one of three phases to ground. The single phase-to-ground fault on LV is shown in Figure 8.14a, the relation between sequence networks under such fault is shown in Figure 8.14b as well as the impedances of each sequence network. The total short-circuit current is

$$I_{+T} = I_{-T} = I_{0T} = \frac{100 \cdot I_r}{0.53 + 0.53 + 1.04} = 47.62 I_r$$

Where $I_r = 25 \times 10^3 / (\sqrt{3} \times 28) = 515.5A$, LV rated current. The short-circuit currents in LV winding are

$$I_+ = I_- = \frac{0.53 \times 47.6 I_r}{8.56 + 0.05} = 2.93 I_r; \ I_0 = \frac{1.04 \times 47.6 I_r}{0.87 + 3 \times 4.78} = 3.26 I_r$$

The total short-circuit current in each winding is

$$I_3 = I_+ + I_- + I_0 = 9.12 I_r; \ I_2 = I_1 = -I_+ + I_0 = 0.33 I_r$$

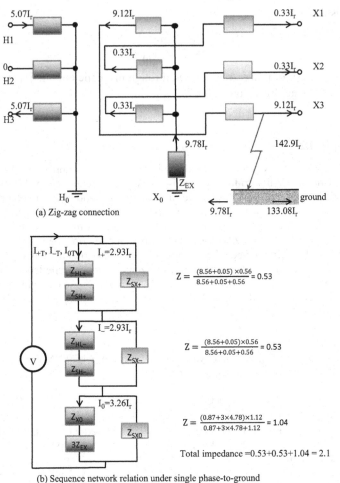

FIGURE 8.14   Single phase-to-ground fault currents.

The current in neutral is

$$I_3 = 3I_0 = 9.78I_r$$

HV winding currents are calculated based on

$$I_{HV} = \frac{I_{zig} \cdot N_{zig} - I_{zag} \cdot N_{zag}}{N_{HV}}$$

$$I_{H1} = \frac{9.12I_r \times 110 - 0.33I_r \times 110}{1466} = 340\,(A)$$

$$I_{H1} = \frac{340}{67} I_R = 5.07I_R; I_{H2} = 0; I_{H3} = 5.07I_R$$

Where $I_R = 67$ A, HV winding rated current.

2. A 45 MVA unit has three windings, HV winding voltage is 245 kV L-L grounded wye, XV winding has 30 MVA, 13.8 kV L-L grounded wye, and YV winding has 15 MVA, 4.16 kV L-L grounded wye. The impedances discussed here are in percentages. The positive sequence impedances at 45 MVA are

$$Z_{HX+}\% = 10.1; \ Z_{HY+}\% = 23.73; \ Z_{XY+}\% = 36.93$$

The impedance of each winding is

$$Z_{H+}\% = (10.1 + 23.73 - 36.93)/2 = -1.55$$
$$Z_{X+}\% = (10.1 + 36.93 - 23.73)/2 = 11.65$$
$$Z_{Y+}\% = (23.73 + 36.93 - 10.1)/2 = 25.28$$

The exciting current = 0.081%. The exciting impedance = 100/0.081 = 1235%, it is so high that it can be treated as infinite. For zero sequence impedances, four tests have been conducted at 45 MVA; they are listed in Table 8.2. From these test results, the zero sequence impedance of each branch is

$$Z_{E0}\% = \sqrt{Z_{10}(Z_{20} - Z_{40})} = \sqrt{49.13(60.5 - 8.39)} = 50.6$$
$$Z_{H0}\% = Z_{10} - Z_{E0} = 49.13 - 50.6 = -1.47$$
$$Z_{X0}\% = Z_{20} - Z_{E0} = 60.5 - 50.6 = 9.9$$
$$Z_{Y0}\% = Z_{30} - Z_{E0} = 72.1 - 50.6 = 21.5$$

The impedance networks are shown in Figure 8.15; here the transformer impedances are assumed reactive. When XV winding's neutral is grounded through Z_{Xext} = 5 Ω resistor, YV winding neutral is grounded through Z_{Yext} = 2Ω resistor. The zero sequence impedance network is shown in Figure 8.15c. The percentages of these two resistances seen from the HV side are

$$Z_{Xext}\% = \left(\frac{245}{13.8}\right)^2 \cdot \frac{3 \times 5 \times 45}{245^2} = 354\%; \ Z_{Yext}\% = \left(\frac{245}{4.16}\right)^2 \cdot \frac{3 \times 2 \times 45}{245^2} = 1560\%$$

The zero sequence exciting impedance is quite reduced compared to positive sequence exciting impedance. If there is additional delta winding, the shunt impedance is reduced further.

TABLE 8.2
Zero Sequence Impedances Tested

Short	Open	Voltage Applied	Tested Results (%)
H1, H2, H3	X1, X2, X3, Y1, Y2, Y3	H123 to H0	$Z_{10} = Z_{H0} + Z_{E0} = 49.13$
X1, X2, X3	H1, H2, H3, Y1, Y2, Y3	X123 to X0	$Z_{20} = Z_{X0} + Z_{E0} = 60.5$
Y1, Y2, Y3	H1, H2, H3, X1, X2, X3	Y123 to Y0	$Z_{30} = Z_{Y0} + Z_{E0} = 72.1$
X1, X2, X3, and H1, H2, H3 to H0	Y1, Y2, Y3	X123 to X0	$Z_{40} = Z_{X0} + Z_{E0}/Z_{H0} = 8.9$

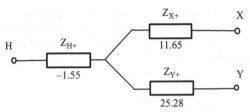

(a) Positive sequence impedance network

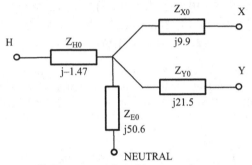

(b) Zero sequence impedance network

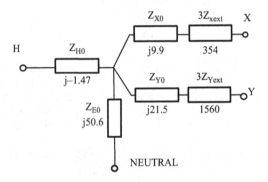

(c) Zero sequence impedance network including neutral resistances

FIGURE 8.15   Three-winding unit impedance networks.

8.7 SHORT-CIRCUIT FORCES ON LEADS

Per IEC [6], thrust forces acting on the lead exits of a high current winding is assumed conventionally to be equal to product of mean hoop compressive stress at the winding (in kN/mm^2) and cross-sectional area of the lead (in mm^2). Where two cable leads are parallel to each other, as shown in Figure 8.16, the short-circuit forces acting on the leads should be checked since it could deform the leads if the short-circuit current is high enough. The force is

$$F = \frac{2\times10^{-7}\times\left(I_1\cdot I_2\right)\times L\times K^2}{r} \tag{8.30}$$

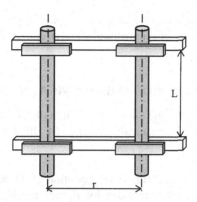

FIGURE 8.16 Paralleled leads clamping structure.

where F is the short-circuit force on the leads, N; I_1, I_2 are rms values of the short-circuit current, A; L is the length of the lead between two clamp bars, m; r the space between two leads, m. If the unsupported length (L) is standardized and the maximum force is known, then space (r) to withstand short-circuit force can be known from this equation.

Example 8.3

The unit LV line has 2000 A current at base rating. The impedance at base rating is Z % = 11.6%. Asymmetrical factor is $K = \sqrt{2} \times 1.93$. Two LV leads are 19 mm (0.75 inch) apart and parallel to it. The distance between two tightening points is 610 mm (2 Feet).

$$F = \frac{2 \times 10^{-7} \times (2000 \times 2000) \times (610 \times 10^{-3}) \times (\sqrt{2} \times 1.93)^2}{(19 \times 10^{-3}) \times (0.116)^2} = 14.2 \, (kN)$$

8.8 THERMAL CAPABILITY OF WITHSTANDING SHORT CIRCUITS

The duration of short-circuit is usually 2 seconds for step-up and step-down power transformers. The maximum permissible average temperature of conductor during it is 250°C for copper, 200°C for aluminum based on IEC60076-5 and IEEE C57.12.00. This duration is unimportant for mechanical stressing on winding or clamp structure since maximum forces occur at first cycle after the fault occurs, then decrease to less than one third after a few half cycles. However, the duration of the fault is important from the winding thermal capability. The highest average temperature of the winding during short-circuit, Θ_1, is calculated as follows:

$$\Theta_1 = \Theta_0 + \frac{2(\Theta_0 + 235)}{\frac{106000}{J^2 t} - 1} \, for \, copper; \quad \Theta_1 = \Theta_0 + \frac{2(\Theta_0 + 225)}{\frac{45700}{J^2 t} - 1} \, for \, aluminum \quad (8.31)$$

Where Θ_0 is the initial temperature of the winding, °C; J is the short-circuit current density based on rms value of symmetrical short-circuit current, A/m²; t is the duration of short circuit, second.

8.9 MEASURES FOR ROBUST MECHANICAL STRUCTURE

Two ways exist to improve the short-circuit strength of transformers. The first is making winding itself mechanically strong by implementing the following measures:

- Reduce current density. Based on Equation (8.1), short-circuit forces will decrease with reduced current density. However, lower current density means more material and larger overall dimension, which is an uneconomical design. It is unwise to pursue this method without exhausting others.
- Use epoxy continuous transposition cable or epoxy CTC. On the top of a polyvinyl acetyl enamel coating of each strand as insulation, there is a final coating of thermos-setting epoxy resin. The epoxy curing occurs at ~120°C for 24 hours, which is oven temperature to get winding dry; it holds all strands together so tightly at normal temperature that the bonding strength of the whole cable is comparable to a massive bar of equivalent cross-sectional area. Regarding the change of the epoxy bonding sheer strength with temperature, some test results show that the bonding sheer strength at ~125°C is about 50% of that strength at 25°C, while other results show that there is no significant decay in the strength with temperature increase, provided that the curing process has been performed properly [1, 2]. A 50% bonding sheer strength at 120°C is usually accepted in the industry.
- Increase yield strength of winding conductor. Generally speaking, the harder the conductor is, the higher its strengths are. However, the harder conductor makes it hard to wind, thus the quality of the winding may suffer.
- Increase winding heights. Based on Equations (8.10) and (8.21), doing so can reduce both radial and axial forces. However, sometimes for large units the winding heights may be limited by shipment height.
- Place compensation gap in LV winding to correspond to the tapping area of HV winding in order to balance ampere-turns between the windings. This way, the radial leakage flux can be reduced by ~50% compared to the case of no compensation gap. This reduction helps to reduce axial force, as well as local heating caused by such radial flux. However, the compensation gap in LV winding creates corners facing HV winding, and the dielectric stress of these corners may exceed the withstand electric strength. In the case that the compensation gap needed is so big that it causes high electric corner stresses of winding, one big compensation gap can be divided into several small gaps by using several thicker radial spacers to thin out sections. This way axial forces are reduced without sacrificing much insulation strength.
- Clamping windings with a higher initial pressure, for example, 5~7 N/mm², to make windings rigid enough in their entire service life. Keeping winding under certain pressure in whole service life is critical for winding conductor to resist

movement under short-circuit forces. The clamping pressures are affected by temperature and moisture content of the windings, its change with temperature is caused by the thermal expansions of the cellulose insulations in the windings, the clamping pressures can be deviated easily by more than 10% from the designed value above for hot winding and below for cold winding. This explains the phenomenon of clamping pressure decrease after cold oil impregnation. In the service the winding temperature changes with load and ambient temperature. The moisture content level of winding depends on tank moisture seal ability, increase of the moisture content causes dimensional expansion of cellulose insulations, and higher clamping pressure as the result. As the consequence from changes of temperature and moisture content, the clamping pressures on the windings are not constant in the whole service life, and decay with insulation aging, one option to keep the windings clamped under certain pressure in their entire service life is to apply high enough pressure the first time.

- Using more or wider-than-usual radial spacers to make the conductor span between the spacers smaller, as the result, the axial bending stress on conductor is reduced. The winding gradient may go up and needs to be checked.
- For windings for severe duty application, such as a furnace transformer, it is preferred that radial key spacers have an outside notch for insertions of sticks to enhance the winding radial tensile strength. When the outer winding has few turns and a small radial build such as tap winding, outer sticks should also be used.
- The mechanical strength of end keep back rings for layer helical winding is a weak point and needs to be checked.
- Short-circuit forces on pitched winding such as layer and helical windings need more attention. The axial forces at every location on circumference of the windings are different. Most calculation programs of short-circuit force don't consider such difference, they treat all windings as disc windings, which has no pitch. The effects of the pitch may be calculated by proper displacement of one winding, for example LV winding, to check forces against strengths.
- When nature frequencies of winding are close to power frequency or twice that of power frequency, a resonance occurs, it could bring in more forces to windings. For larger units, the nature frequency of windings should be checked to avoid such closeness. Generally speaking, radial resonance frequency of a winding is higher than twice that of power frequency, so resonance seldom happens. Axial resonance frequency of a winding depends on properties of its conductor and insulation materials, as well as on its clamping forces. The higher the clamping force, the more rigid the winding, and the higher the resonance frequency [7].

The second way is to reduce short-circuit current by increasing impedances. The impedance is increased, for example, by increasing the distance between windings. Doing so will increase the overall dimension and cost, too. Another way to increase impedances between HV or LV windings and delta tertiary winding is to insert a current limiting reactor (CLR) in the tertiary winding circuit; the selection criteria of the current limiting reactor is discussed in reference [8].

Ideally, the current limiting reactor (CLR) should have no iron circuit because the iron circuit gives a nonlinear saturation-type characteristic; in an over-current situation, its reactance tends to decline, which weakens its short-circuit performance [4]. On the other hand, the shield is needed around CLR's coils to prevent the coil flux from entering into the tank and thus reducing loss and heating. In the case of magnetic shield (core steel shield), if the shield cross-sectional area is reduced, the flux density under normal rated condition increases and the chance to saturate under short-circuit current is greater, thus bringing a greater reactance reduction. In the case of electromagnetic shield (copper or aluminum shield), the reactance is not affected as much as the case of magnetic shield; however, the cost is higher. The design should reflect an balance between the reactor size/cost and its performance.

Like transformer windings, the reactors are subjected to a large current under short-circuit conditions. Since each phase has only one winding, there is no significant axial unbalance as can be experienced in a transformer, so there is no major end force on winding supports. An axial compressive force and an outward bursting force are in the winding; the latter is resisted by the yield strength of the copper, which is usually able to meet this force, but the winding must be adequately braced to prevent any tendency for it to unwind. Since the reactor windings normally have fewer turns than transformer windings, this aspect often requires more careful consideration.

A transformer as whole has a mechanical strength determined by the strength of its weakest component. The strength of a transformer for a single fault event may be considerably higher than that for a series of faults, since weakening of the winding and axial displacement could be progressive [4]. Based on these, it is common practice to give enough margins between short-circuit stresses and its corresponding strengths. IEC 60076-5 lists the minimum strengths required for each type of stress [6]. The calculations of each stress are listed here; the deductions of the calculation equations can be found in mechanical engineering books.

8.10 COMPRESSIVE STRESS ON RADIAL SPACER

$$\sigma_{spr} = F_{spr}/\left(N_{spr} \cdot W_{spr} \cdot RB\right) \tag{8.32}$$

Where σ_{spr} is stress at the radial spacer, N_{spr} is number of spacers per circle, W_{spr} is width of one spacer, RB is copper radial build of winding; F_{spr} is maximum axial compressive force on a section. It is suggested that σ_{spr} is equal to or less than 80 MPa and 120 MPa for paper-wrapped conductor and paperless conductor respectively.

8.11 AXIAL BENDING STRESS ON CONDUCTOR

$$\sigma_{axial} = \frac{F_{axial}}{2 \cdot b} \cdot \left(\frac{L}{h}\right)^2 \tag{8.33}$$

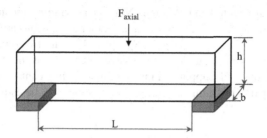

FIGURE 8.17   Sketch for axial stress calculation.

Where h is axial dimension of one strand conductor, m; b is radial dimension of one strand conductor, m; F_{axial} is distributed axial force on one strand, N/m, as shown in Figure 8.17. It is suggested $\sigma_{axial} \leq 0.9R_{p0.2}$, where $R_{p0.2}$ is proof stress producing 0.2% permanent strain [6]. From Equation (8.33), in order to reduce the axial stress, reducing the span L by increasing the number of spacers per circle is the most effective way, when the conductor size cannot be changed by other reasons such as load loss.

8.12 TILTING FORCE

Another phenomenon which could occur under high axial force is the tilt of conductors between the spacers, as shown in Figure 8.8. Calculation of the critical tilt force is given in reference [6]. When axial force is higher than the critical tilt force, the tilting could happen, it is necessary during the design stage to ensure that strand conductors or non-epoxy CTC do not tilt under short-circuit force as well as under clamping forces. Epoxy-bonded CTC doesn't tilt after the epoxy is cured; that is, such CTC strands are extremely resistant to tilting under short-circuit forces in service. Attention is needed to possible conductor tilt before epoxy is cured in the oven. Inadequate conductor size, loose wound coil, high pressure for oven baking can result in CTC strands tilt.

8.13 HOOP STRESS

Hoop stress is tensile or compressive, depending on whether the pressure acts outwards or inwards radially. The value of the stress is estimated per Equation (8.18), which is represented again as follows:

$$\sigma_{sc.rad} = 4.76 \times \frac{P_{DC@75}}{H_{wdg}} \cdot F_{s\,axial} \cdot \left(r\right)^2 \left(\sqrt{2} \cdot k\right)^2 \tag{8.34}$$

This equation is for both compressive and tensile stresses. The compressive stress on an inner winding and the tensile stress on an outer winding are also expressed as

$$\sigma_{inner\ radial} = \frac{F_{radial}}{2 \cdot h} \cdot \left(\frac{L}{b}\right)^2 \; ; \; \sigma_{outer\ radial} = \frac{F_{radial} \cdot D}{2 \cdot a} \tag{8.35}$$

where F_{radial} is distributed radial force, N/m; D is winding mean diameter, m, a ($\approx h \times b$) is conductor cross-sectional area, m^2; other dimensions are shown in Figure 8.17. It is suggested that $\sigma_{outer, radial} \leq 0.9R_{p0.2}$ [6]. There are two types of phenomena happening to inner winding under compressive forces: one is the free buckling phenomenon, where the stick supports at inside of the winding cannot resist it at all. The critical stress for free buckling, $\sigma_{freebuckling\ cri}$ in N/mm^2, may be estimated by the following equation:

$$\sigma_{freebuckling\ cri} = \frac{E_0 \cdot b^2}{4 \cdot R^2} \qquad (8.36)$$

Where E_0 is conductor Young's modulus or modulus of elasticity, for copper $E_0 = 1.10 \times 10^5$ N/mm^2, b is conductor radial dimension(thickness), m, R is the mean winding radius, m. This equation doesn't consider the conductor proof stress, which is a major factor in resisting free buckling. In reality, the strength to resist free buckling is a more complicated issue because several factors such as non-homogeneous nature of the windings, manufacturing qualities like winding tightness and clamping uniformity play a role. It is suggested to use a simple method [6], and for windings made by regular strands or non-bonded CTCs, $\sigma_{freebuckling,\ cri} \leq 0.35R_{p0.2}$, for winding made by epoxy-bonded strands or CTCs, $\sigma_{freebuckling\ cri} \leq 0.6R_{p0.2}$. Another phenomenon under compressive force is forced buckling, where the inside sticks resist the deformation, the conductors are bended between the sticks. It is suggested that $\sigma_{forcedbuckling\ cri} \leq 0.9R_{p0.2}$. The critical stress for forced buckling in N/mm^2 may be estimated by [9]:

$$\sigma_{forcedbuckling\ cri} = \frac{E_t}{12} \cdot \left(\frac{b}{R}\right)^2 \left(\frac{N^2}{4} - 1\right) \qquad (8.37)$$

where N is number of radial supports per circle; E_t is tangential modulus of elasticity in N/m^2

$$E_t = \frac{E_0}{\left[1 + \gamma \left(\frac{\sigma_p}{\sigma_0}\right)^m\right]}; \ \sigma_0 = \left(\frac{k}{E_0 \cdot \varepsilon_p}\right)^{1/m} \sigma_p^{m+1/m} \qquad (8.38)$$

For copper, $k = 3/7$; $m = 11.6$, $\gamma = k\,(m + 1) = 5.4$. Where $\varepsilon_p = 0.002(0.2\%)$ is residual strain, σ_p is copper proof stress.

Example 8.4

The copper wire of an inner winding has proof stress of 200 N/mm^2 at 0.2% strain. The winding mean diameter is 1015 mm, the wire size is 10 mm axial by 1.65 mm radial. The winding has 20 conductors in the radial build and 20 inside axial sticks. Its σ_0 is

$$\sigma_0 = \left(\frac{3/7}{1.1 \times 10^5 \times 0.002}\right)^{1/11.6} 200^{(12.6/11.6)} \approx 184\left(N/mm^2\right)$$

Under 200 N/mm² stress, the tangential modulus of elasticity E_t is

$$E_t = \frac{110000}{1 + 5.4 \times \left(\dfrac{200}{184}\right)^{11.6}} \approx 7234 \left(N / mm^2\right)$$

The critical stress for forced buckling is

$$\sigma_{forcedbuclking\ cri} = \frac{7234}{12} \times \left(\frac{1.65 \times 20}{1015 / 2}\right)^2 \left(\frac{20^2}{4} - 1\right) \approx 252 \quad \left(N / mm^2\right)$$

The critical stress for free buckling is

$$\sigma_{freebuckling\ cri} = \frac{110000 \times \left(1.65 \times 20\right)^2}{4 \times \left(1015 / 2\right)^2} \approx 116 \left(N / mm^2\right)$$

As can be seen, the free buckling critical stress is lower than forced buckling critical stress. This is the reason that free buckling happens under lower forces. Windings shall be designed to be self-supported, so they have enough strength to resist deformities caused by free buckling.

REFERENCES

1. Giorgio Bertagnolli, *The ABB Approach to Short-Circuit Duty of Power Transformers*, Third revised edition, ABB Management Services Ltd Transformers, Switzerland, 2007.
2. Daniel Hermann Geißler, *Short-Circuit Withstand Capability of Power Transformers*, Cuvillier Verlag Göttingen, 2017.
3. M. Waters, *The Short-Circuit Strength of Power Transformers*, MacDonald, London, 1966.
4. Martin J. Heathcote, *The J&P Transformer Book*, 13th edition, Elsevier, Newnes, Amsterdam, et al., 2007.
5. IEC 60076-8, *Power Transformer-Application Guide*, 1st edition, 1997.
6. IEC 60076-5, *Power Transformers Part 5: Ability to Withstand Short Circuit*, 3rd edition, 2006.
7. Xie Yucheng, *Power Transformer Handbook*, China Machine Press, Beijing, 2009.
8. M. J. Lantz, et al., *Fault-Current Limitation for Large Transformers Using Reactors in Tertiary Winding*, IEEE Transaction-PAS, June 1964.
9. Robert M. Del Vecchio, et al., *Transformer Design Principles with Application to Core-Form Power Transformers*, 2nd edition, CRC Press/Taylor & Francis Group, Boca Raton, London, New York, 2010.
10. Working group 12.19, *The Short-Circuit Performance of Power Transformer*, Cigre, August 2002.

9 Sound Levels

Noise is one of those environmental contaminants which should be reduced. A noise is defined as a sound which is undesired by the recipient. The annoyance value of transformer noise is roughly proportional to its apparent loudness. Loudness is a subjective sensation dependent on the characteristics of the human ear. Research shows that the loudness is a reasonably well-defined function of its sound pressure and frequency. Sounds with frequency between approximately 16 Hz and 16 kHz are audible to most human observers. Before the source, transmission path of transformer sound, methods for its reduction and measurement are discussed; the definitions of three sound levels are introduced.

Three popular measurements of sound level are sound pressure, sound power and sound intensity. Sound source radiates sound power; it is a unique physical property of the sound source and independent of any external factors such as distance from receiver and environment in which the sound waves travel. What we hear is sound pressure, caused by the sound power emitted from the source, and dependent on the distance from the source and the environment.

Sound power is the rate at which energy is radiated (energy per unit time). Sound intensity is the rate of energy at a point, that is, through a unit area. Sound intensity is a vector quantity. Sound pressure is a scalar equivalent quantity having only magnitude. Normal microphones are capable only of measuring sound pressure, this is sufficient for majority of transformer sound measurement situations.

Sound level measurement is defined as the weighted sound pressure level. Historically, A, B and C weighting networks were specified to simulate the response of the ear at low, medium and high noise levels respectively. However, extensive tests have shown that in many cases the A-weighted noise level correlates best with subjective noise ratings and is now widely used. Although C weighting is retained in more comprehensive meter, B weighting has fallen into disuse [1].

The major sources of sound radiated by transformers are

- No-load sound or core sound. It is produced by magnetostriction of core steel laminations themselves, and magneto-motive force between laminations. This sound has frequencies mainly of 120, 240, 360 and 480 Hz when the transformer operates at 60 Hz (100, 200, 300 and 400 Hz when it operates at 50 Hz). These sounds are low-frequency and tonal in nature, which may propagate farther with less attenuation. When being energized, the transformer produces no-load sound regardless of whether it has load or not.
- Load sound. This is the sound emitted by a loaded transformer in addition to its no-load sound. The load sound is the result of vibrations of windings, tank wall and magnetic shunts under electromagnetic forces which are produced by interactions between leakage flux and current in the winding. The load sound has a frequency of 120 Hz when the unit works at 60 Hz (100 Hz when it works at 50 Hz).

- Cooling equipment such as a fan and pump also produce sound during their operation. The frequencies of the cooling equipment's sound dominate at lower and higher ends of the frequency spectrum.

The sounds have the following features in general [2]. First, according to the laws of acoustics, the volume of sound pressure $L_{p(d)}$ decreases with the distance d which is from the assumed point source, i.e. the center of the equivalent hemisphere, to the measurement point as

$$L_{p(d)} \approx L_{p(3m)} - 20\log\left(\frac{d}{3}\right) \tag{9.1}$$

where d is distance in meters and $L_{p(3m)}$ is sound pressure measured at 3 meters. Based on this equation, when the distance d is doubled, the sound pressure level decreases by 6 dB. The real measurements [3,4] show that this relation sometimes can be found in the far field (30 m away from the unit), not in the intermediate region. Because different harmonics attenuate at different rates, these rates may be different in different transformers. Another feature is that the sound pressure varies with the frequency as

$$L_{p(f)} \approx L_{p(60)} + 20\log\left(\frac{f}{60}\right) \tag{9.2}$$

where $L_{p(60)}$ is sound pressure level at 60 Hz. In the following discussion, when "sound level" is used, it means sound pressure level.

9.1 NO-LOAD SOUND

Magneto-motive force and magnetostriction contribute to no-load sound or core sound, they are discussed next.

9.1.1 Magneto-Motive Force

Where fluxes jump from one lamination to an adjacent lamination, magneto-motive forces are generated; this type of flux is sometimes called cross-flux. The magneto-motive forces make the laminations to vibrate perpendicular to the lamination plane at twice the supply frequency and to produce sound [3]. In modern transformers, the cross-flux is relatively small except at the corners or at the joints, where air gaps exist in the magnetic path. Example 9.1 shows the effect of magneto-motive force on the sound.

Example 9.1

Two units of the same design, one core is machine stacked, while another core is hand stacked; the no-load sound level of each unit is listed in Table 9.1. The gaps at corners and joints of hand stacked core are found bigger compared to machine stacked core, these bring about 4 dB increase in sound pressure level.

TABLE 9.1
Sound Level Comparison

		Exciting Current (%)	
	At base MVA (dB)	At 100% voltage	At 110% voltage
Unit 1, hand stacked core	68.8	0.17	0.93
Unit 2, machine stacked core	64.3	0.15	0.71

Reducing the number and size of air gaps in the core can reduce noise produced by cross-flux or magneto-motive force. Step lap core is one measure used widely for such purpose; its sound level is lower than butt lap core. The cross-flux also exists where there is an appreciable difference in permeability between adjacent laminations, laminations in core are not laid flat or laminations have non-uniform thickness. The magneto-motive force per unit cross-sectional area, F_m, is

$$F_m = \frac{B_{peak}^2}{4\mu_0}\left(1 + \cos 2\omega t\right)$$

(9.3)

where B_{peak} is peak value of flux density in the gap, μ_0 ($= 4\pi \times 10^{-7}$ H/m) is permeability of free space, ω is the fundamental angular frequency and t is time. In real cores, the laminations are not laid flat as they are not clamped together all over the entire surface; residual gaps between the laminations and at joints are unavoidable. Magneto-motive forces across these air gaps could set relative transverse motion of the laminations, because the laminations adjacent to the gaps or bridging the gaps, and laminations edges at joints possess freedom of motion, both produce sound. The unit, having higher-than-expected tested exciting currents compared with other identical transformers, may possibly be tested higher sound levels. When a core has more gaps, it needs more exciting power or higher exciting current to drive flux through the magnetic circuit; this interprets partly why exciting current is a measure of the steel quality and size of gaps in core. Besides higher exciting current, more gaps give higher chance of lamination vibrations. In Example 9.1, the exciting currents of both units are higher than the calculated one (0.127%), but not much higher, maybe because of the small core.

9.1.2 MAGNETOSTRICTION

In 1842, James Prescott Joule discovered that an iron rod changes its length when subjected to a magnetic field whose direction was parallel to the axis of the rod. Later, it was found that all ferromagnetic materials have such a property. The ferromagnetic materials consist of magnetic domains, which are tiny regions of uniform magnetic polarization. When an external magnetic field is applied, the boundaries between the domains shift or the domains rotate, both of which cause a dimension change. This phenomenon is called magnetostriction, and the force making its occurrence is called magnetostriction force. This change not only causes longitudinal vibrations in the plane of the lamination but also results in out-of-plane displacement of the lamination. Studies show that core steel in a magnetic field changes its dimension as follows [5]:

$$\Delta l = l \cdot \sum_{\upsilon=1}^{n} K_v \cdot B^{2\upsilon}$$

(9.4)

where B is the instantaneous value of flux density, K_v is coefficient which depends on level of magnetization type of lamination material and its treatment. With the increase of exponent (order number v), the coefficients K_v usually decrease. The magnetostriction force, F, is given by

$$F = E \cdot A \cdot \sum_{\upsilon=1}^{n} K_{\upsilon} \cdot B^{2\upsilon}$$

(9.5)

where E is the modulus of elasticity of core steel in the direction of force, and A is the cross sectional area of a lamination sheet. The equation indicates that the magnetostriction force varies with time and contains even harmonics of the power frequency (120, 240, 360, ...). This force causes core steel to vibrate and produces noise with the same frequencies. The amplitude of core vibration and noise could increase significantly if the mechanical natural frequency of core is close to 120 Hz or its multiples, the resonance occurs.

Magnetostriction is sensitive to mechanical stress at core laminations, generally speaking, the higher the compressive stress that is exerted on the lamination, the higher the magnetostriction, the greater the contribution to the core noise. Efforts to obtain flat lamination should be tried as clamping of wavy lamination in a core will result in increased stress as well as increased noise level. On the other hand, if no pressure is exerted on the core, the gaps between laminations become larger, and the vibration caused by magneto-motive force and its consequent sound increase. As shown in some researches, the vibrations can be reduced to some extent by tightly clamping the core, especially the yokes [6], to prevent vibration perpendicular to the plane of lamination.

Investigation on effect of silicon content of core steel on magnetostriction found that steel having 6% to 6.5% silicon is least affected by magnetostriction. However, the brittleness of such steel makes its practical use difficult. The sound level caused by magnetostriction, $L_{p, core, stri}$ may be written as follows:

$$L_{p,core,stri} = K_1 + 6.67 \cdot \log_{10} W$$

(9.6)

where K_1 is a coefficient and W is core weight. Considering other factors such as flux density, resonance and harmonic vibrations, and vibrations caused by magneto-motive force, the core sound level may be presented as

$$L_{p,core} = K_{weight} \cdot \log_{10} W + K_{flux} \cdot \log_{10} (B) + K_{resonance} + K_{radiator\ effect}$$

(9.7)

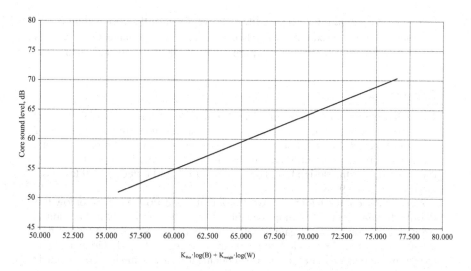

FIGURE 9.1 Tendency of core sound level.

where $L_{p,core}$ is sound pressure level in dB, W is core weight, K_{weight} is weight coefficient; B is flux density, K_{flux} is flux density coefficient and $K_{resonance}$ is contribution from core resonance. Like any mechanical structure, transformer core structure has several natural frequencies. If the core structure vibrates with a frequency near one of these natural frequencies, resonance occurs, and it could increase the sound level significantly. $K_{radiator\ effect}$ takes locations of radiators into account because their locations affect measured sound level. Figure 9.1 shows the tendency of tested core sound level; all test units have laser-scribed core steel, and the tests were conducted per IEEE C57.12.90.

9.1.3 Transmission

Core vibration is transmitted to the tank through the core supports, which is structure-borne transmission, and through the oil, which is oil-borne transmission. Reduction of the noise level could be achieved by reducing the amounts transmitted by these two channels.

9.1.4 Abatement Techniques

The techniques of core sound level reduction are divided based on weakening the sound source, or on isolating the sources [3, 5, 7, 8, 9] from the recipient.

9.1.4.1. To Weaken the Sources

The following methods listed here are used in practice.

- Reduce flux density in the core to reduce its sound as shown by Equation (9.7). It is perhaps the simplest way to reduce the sound level. However, there exist certain extents, beyond which further reduction in flux density doesn't help to reduce the sound level because the negative effect of the core weight is notice-able, as shown in Example 9.2. Also, a bigger core is not an economical design since more core steel and copper are required for the same performances.

Example 9.2

Two units are required to have sound levels much lower than standard limits. Unit 1 has core flux density of 1.4 Tesla, about 20% lower than normal value; unit 2 has core flux density of 1.1 Tesla, which is extremely low; and larger core and coils as a consequence. Both core materials are the same. The sound levels are estimated during the design based on Equation (9.7) and tested in final test. Both are listed in Table 9.2.

As can be seen, the larger core weight of unit 2 eats up the benefit of low flux density. Further, unit 2 has more weights of core steel and copper than unit 1. Another thing that needs to gain attention is, when units operate with lower than 1.6 to 1.5 Tesla, the load sound from windings may easily be a dominant factor in transformer total sound. In such a case, reducing the core sound level doesn't help lower the total sound, as shown in Example 9.3.

- Hi-B scribed material gives a reduction of 2~7 dB compared to non-Hi-B grade.
- The step lap joint can reduce core sound by 4~5 dB compared to the butt lap joint. The reason is as mentioned before: the step lap joints reduce the number and size of air gaps in the core, in turn reducing the noise produced by mag-neto-motive force. Using a multi-step lap joint instead of a single-step lap can achieve 7~8 dB reduction between 1.4 and 1.5 Tesla, while at 1.7 Tesla the reduction is negligible. It may imply that at 1.7 Tesla, the sound by magneto-striction may be predominant. To achieve a significant improvement on the reduction, the multi-step lap has to consist of 3~4 steps; a further increase in the number of steps seems not have much of an influence.
- Placing special rubber sheets between core laminations to absorb vibration. It should be noticed that the core area will be reduced due to the insertion of the rubber sheets. As a consequence, the core flux density will increase as will the noise level to diminish the effort. This measure has to be carefully and thor-oughly evaluated before implementing. Also, this method reduces the vibration caused by magneto-motive force, not by magnetostriction. When the noise is caused mainly by magneto-motive force, this method may help.

TABLE 9.2
Comparison of Core Sound Level Between Unit 1 and Unit 2

	Flux Density (Tesla)	Core Sound Estimated (dB)	Core Sound Tested (dB)	Core Weight (kg/MVA)	Copper Weight (kg/MVA)
Unit 1	1.4	58.0	55.8	694	306
Unit 2	1.1	58.2	55.04	846	337

- Avoid core, tank and stiffener resonance. This resonance could increase sound level by 2~10 dB [10].
- Reduce cross-flux to reduce magneto-motive force. Using a flat steel sheet, keeping the flux distortion to a minimum may reduce sound level by 1~2 dB.

9.1.4.2 To Reduce the Transmission

- Reduce the structure-borne vibration transmission. The sound radiated by the transformer tank is generated by both oil vibration coupling to the tank and the structure-borne vibration of the tank which is rigidly attached to the core. In general, the structure-borne vibration can be reduced by changing stiffness of the core supporting structure or by using vibration absorber, or by increasing its mass. For example, core assembly is normally rigidly mounted to the tank base and walls to restrain core movement during shipping. When possible, placing the core assembly on springs or inserting anti-vibration pads between the core assembly and tank base and wall could damp tank wall vibration transmitted through the core support structure, as well as the related sound. If possible, the core clamping frame may not be rigidly and directly tied to any part of the tank wall at any point. Such method can give a sound level reduction of 1~3 dB [10,11]. This method mitigates the transmission of the core vibration to the tank base and wall. To achieve good vibration isolation, two factors have to be considered. One factor is that the mechanical impedance of isolating members, i.e. springs, transport bracing, oil between core and tank, should be as low as possible. The other factor is that mechanical impedance of the members that shall be isolated from each other, i.e. the core and tank, should be as high as possible.
- Reduce the oil-borne vibration transmission. Since the oil is relatively incompressible, the proportion of vibrational power as well as the sound transmitted through it may be appreciable. Mounting a vibration absorber on the tank wall may help. Exterior sound panels between tank stiffeners are used and may reduce sound level by 6~10 dB. However, it reduces the cooling capabilities of the tank wall so that additional cooling equipment is needed. There may be a maintenance issue when these panels get wet.
- Bad quality of yoke clamping can give an extra 10 dB to the sound level. Effort on making laminations as straight as possible, not waving, helps reducing the magneto-motive and magnetostrictive forces and the sound level produced.
- Increase tank wall mass (impedance) to reduce vibration by filling sand in hollow braces on the wall. This can reduce the sound level by about 2 dB.
- The double tank wall and the inner and outer walls are isolated from each other to eliminate structure-borne vibrations, also, suitable sound-absorbent wool is placed between the two walls, the noise level reduction is about 6~10 dB with sound panel, 5~8 dB without the sound panel [10]. The disadvantage of the double tank wall is that the secondary wall substantially increases the unit weight and size; this creates shipping and handling difficulties. Further, maintenance of the double tank wall unit is more difficult because access to the primary wall is very limited. Oil leaks, for example are hard to detect and fix. The tank wall thickness may be increased to improve the sound level.

- Field-installed sound abatement techniques and active sound cancellation techniques are listed in IEEE Std. C57.136. Vegetation such as forest and shrubs are not generally considered an effective noise barrier, although it does have an effect in attenuation noise at frequencies above 2 kHz. However, the psychological effect of vegetation as a barrier between a noise source and an observer should not be overlooked. In many cases if the noise source is not visible, the noise is less noticeable and thus less annoying.

A shunt reactor has a gapped core, the vibration is quite high compared to a transformer due to high magnetic force between every two magnetic packets separated by a non-magnetic gap. The methods for reducing noise level are very stiff structures to eliminate excessive vibration, using a high modulus of elasticity material such as a stone spacer or ceramic blocks as gap material, placing epoxy impregnated polyester material and fiberglass cloth of 2~3 mm between the last leg packet and top and bottom yoke. The material is hardened by the heating during the processing stage.

9.2 LOAD SOUND

9.2.1 SOUND FROM WINDING

The load sound comes from winding. It is produced by Lorentz forces that result from the interaction between leakage flux generated by one current-carrying winding and the current in the conductors of other winding. This force causes vibrations of windings in both axial and radial directions, resulting in acoustic radiations at twice the power frequency.

The measurement and simulation conducted in reference [12] show that the surrounding oil does not affect the axial accelerations of the winding support platform. However, due to the mass-loading effect of the surrounding oil, the radial acceleration amplitudes are nearly halved when compared to the vibrations of winding without oil surrounding; reference [4] gets a similar result. Second, with decrease of the radial surface accelerations of the outmost winding, tank surface vibrations, A-weighted sound power level are greatly reduced. Third, neglecting the axial stiffness of the windings and core supports causes significantly increased axial clamping accelerations and slightly decreased radial coil accelerations, therefore greatly increasing the sound power level. These results indicate that the stiffness has a strong influence on the load sound level.

Investigations have further shown that different winding types and their arrangement related to core influence the load sound [7]. The winding tightness also plays an important role for the load sound as well. Furthermore, the winding's mechanical natural frequencies in the vicinity of twice the power frequency should be avoided; otherwise the resonance of winding would cause an increase up to 5 dB.

From test results, in general the load sound could make the transformer total sound pressure level 2~2.5 dB higher than no-load sound. If the load current contains significant harmonics like rectifier transformers, these additional harmonics are a significant source of load sound [13]. Based on factory test results of 65 transformers with ratings of 60–1000 MVA, 50 Hz operation, the load sound power level, L_{load}, can be roughly estimated by [14].

$$L_{load} \approx 39 + 18 \cdot \log_{10} \frac{S_r}{S_p} \qquad (9.8)$$

where L_{load} is the A-weighted sound power level of transformer at rated current, rated frequency (50 Hz) and impedance voltage, S_r is rated power in MVA. For autotransformers and three-winding transformers, the equivalent two-winding-rated power, S_r, is used instead of S_r. S_p is the reference power (1 MVA). For a 60 Hz transformer, the sound level increases about 5 dB, so the sound power level is roughly estimated as.

$$L_{load} \approx 44 + 18 \cdot \log_{10} \frac{S_r}{S_p} \qquad (9.9)$$

Test data have shown that actual load sound power levels can vary mostly between +12 dB and −6 dB from the level calculated using the aforementioned equations. It is understandable because different manufacturers use different clamping structure and material. Figure 9.2 shows a general trend of the load sound test result, where P is parameter related to winding copper weight, current density, MVA and impedance. A different load current has different sound power level. A guide to estimate the sound level at different load current is [4]:

$$L_{load} \approx L_{rated} + 40 \cdot \log_{10} \frac{I}{I_N} \qquad (9.10)$$

where L_{load} is A-weighted sound level (dB) at current I, and L_{rated} is A-weighted sound level (dB) at rated current I_N. As can be seen, reducing the current by 50% provides 12 dB reduction in sound level. The equation is valid for the current I in the range of 60~130% of the rated current I_N. The formula does not include magnetrostriction effects caused by leakage flux in magnetic shunts. The load sound from vibration of the tank wall, magnetic shunts and winding dominate the intermediate frequency range, and it has one tone: twice that of the power frequency. These vibration forces are proportional to the square of the load current, the sound is proportional to the fourth power of the current. For transformers operating with flux density lower than 1.5 T to 1.6 T, the load sound becomes more obvious and may easily become the dominant portion of transformer total sound.

9.2.2 SOUND FROM TANK WALL

In order to avoid tank resonance, it should make tank mechanical natural frequency not close to twice of power frequency. Also increasing tank mass and stiffness help the reduction of its vibration.

9.2.3 SOUND FROM MAGNETIC SHUNTS

Magnetic shunts are frequently used in large transformers to reduce eddy current losses generated in metallic parts in the tank as well as the tank wall. At full load, the leakage flux in the shunts may reach to such a level that the sound generated by magnetostriction makes a significant contribution to the total sound level [15]. For a low-sound-level unit, the magnetic shunt's contribution to the sound level has to be carefully considered. Alternatives are to increase the tank size or stainless steel patches in the tank wall to avoid application of the magnetic shunt.

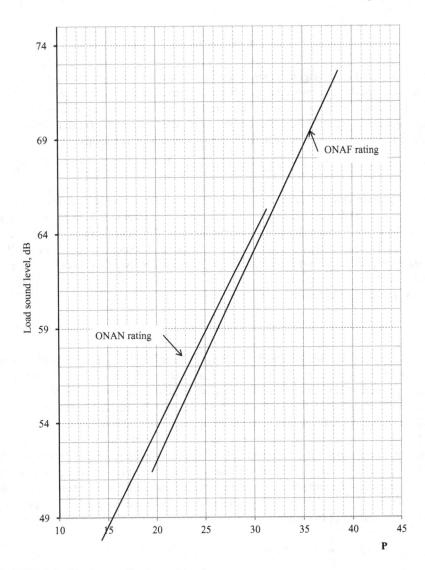

FIGURE 9.2   Tendency of load sound level.

9.2.4 ABATEMENT TECHNIQUES

Several methods for load sound level reduction are listed here as reference.

- Use windings that provide lower magnitudes of leakage flux density. With the same ampere-turn, a taller winding has a lower leakage flux density than a shorter winding. This means that the taller winding has lower impedance than the shorter winding, and low-impedance units usually have lower load sound levels. However, the impedance is usually a guaranteed parameter.

- Make the winding wound tight and clamped tight to increase its stiffness in both axial and radial directions. It is found that winding vibration in a radial direction has no significant effect on load sound, while axial vibration is a major contributor unless it is very large unit. A good, tight winding support structure is essential for a low level of load sound.
- Winding mechanical natural frequencies in the vicinity of twice the power frequency should be avoided.
- Reduce flux density in the tank shunts, damp treatment of the tank and sound enclosures covering the entire tank as discussed before. The tank thickness variations of 1 mm can imply sound power level differences of nearly 5 dB. This rather large discrepancy in sound power can be attributed to the impact of the tank eigen frequencies on sound transmission [16].
- Changes of 20% in Young's modulus of the radial spacer in winding result in sound power level differences of ~1.5 dB [16].

9.3 FAN SOUND

The sound levels of cooling fans dominate at the lower and higher ends of frequency spectrum of the total sound. The influence factors on the fan's sound level are tip speed, number of blades, blade design, number of fans and arrangement of radiators. Typical relations of sound level with different fan speed (RPM) and quantity of fan is shown in Figure 9.3. The highest no-load sound level that both core and fan are in operation, $L_{p,no\text{-}load}$, is the logarithmic sum of the core sound level and the fan sound level.

$$L_{p,no-load} = 10 \cdot \log_{10}\left(10^{0.1 \times L_{p,core}} + 10^{0.1 \times L_{p,fan}}\right) \tag{9.11}$$

Where $L_{p,core}$ and $L_{p,fan}$ are sound levels of core and fan respectively in dB.

9.4 TOTAL SOUND

The total sound pressure level of a transformer, $L_{p,total}$, is the logarithmic sum of its no-load sound pressure level, $L_{p,no\text{-}load}$, and its load sound pressure level, $L_{p,load}$ as follows:

$$L_{p,total} = 10 \cdot \log_{10}\left(10^{0.1 \times L_{p,no-load}} + 10^{0.1 \times L_{p,load}}\right) \tag{9.12}$$

The total sound pressure will never be more than 3 dB higher than its louder individual sound. Also, the uncertainty in the total sound is not greater than the uncertainty in the measurement of any of the contributing sounds. The total sound power level, $L_{w,total}$, is calculated by the similar formula.

$$L_{w,total} = 10 \cdot \log_{10}\left(10^{0.1 \times L_{w,no-load}} + 10^{0.1 \times L_{w,load}}\right) \tag{9.13}$$

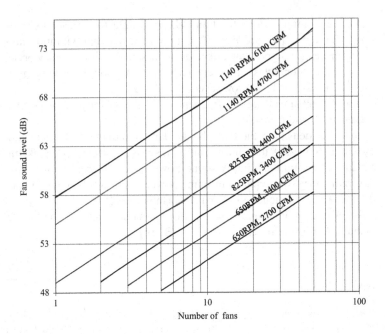

FIGURE 9.3 Sound level of fans with different speed and quantity.

Where $L_{w,no\text{-}load}$ is A-weighted sound power level at rated voltage on open circuit, $L_{w,load}$ is A-weighted sound power level at rated current. Sound power levels can be added or subtracted to provide sound power level for cooling equipment alone when the total sound power level of the transformer and cooler has been obtained, but this must be done logarithmically. For example, the sound power level of the cooling equipment, $L_{w,cooling}$, is determined by

$$L_{w,cooling} = 10 \bullet \log_{10}\left(10^{0.1L_{w,total}} - 10^{0.1L_{w,tx}}\right) \qquad (9.14)$$

Where $L_{w,total}$ is the sound power of the transformer and cooler, $L_{w,tx}$ is the sound power of the transformer alone.

Example 9.3

The tested no-load sound pressure level of a unit is 58.7 dB, the tested load sound pressure is 62.04 dB. The tested total sound pressure is

$$L_{p,total} = 10 \times \log_{10}\left(10^{0.1 \times 58.7} + 10^{0.1 \times 62.04}\right) = 63.7\ dB$$

From the test results, it shows that a very low no-load sound level achieved by very low flux in the core doesn't help the total sound level, because the load sound level becomes a dominant factor.

9.5 SOUND LEVEL MEASUREMENTS

9.5.1 SOUND PRESSURE LEVEL

Sound pressure is a pressure fluctuation in the air which is caused by a sound source; it is this pressure fluctuation which our ear hears and which is measured by the microphone of a sound level meter. The measured value is related to distance from the source, frequency, background noise and sound reflection. Transformer "hum" is of low frequency, fundamentally 100 Hz for a 50 Hz unit or 120 Hz for a 60 Hz unit. The normal range of hearing for a healthy young person extends from approximately 20 Hz to 20 kHz. The definition of the sound pressure level is

$$L_p = 20 \times \log_{10}\left(\frac{p}{p_0}\right) \tag{9.15}$$

where p is measured sound pressure and p_o is a reference pressure of 20 micro-Pascal. The sound pressure is a scalar quantity. A useful aspect of decibel scale gives a better approximation to human perception of relative loudness than the linear pascal scale; this is because our ear responds to sound logarithmically. However, our ear does not respond by the same amount for each frequency; hence, a suitable filter is required to ensure that the microphone measurements truly reflect the sound perceived by the ear. An internationally standardized filter, termed "A weight", addresses this requirement. Measured sound level is actually a combination of transformer and ambient sound pressure levels, and the transformer sound pressure level alone can be obtained after ambient sound and sound reflection corrections; these corrections are stated in [10,17]. The tested average sound pressure level is calculated by

$$L_p = 10 \times \log_{10}\left[\frac{1}{N}\sum_{i=1}^{n} 10^{(L_{pi}/10)}\right] \tag{9.16}$$

where L_{pi} is the sound pressure level measured at the ith location, and N is the total number of sound level measurements. As mentioned previously, the measured values by microphone are affected by background noise, wall sound reflection and the near-field effect, so the measured values should be corrected before calculating the average sound level.

9.5.2 SOUND INTENSITY LEVEL

Sound intensity is defined as the rate of energy flow per unit area and is measured in Watts/m². It is a vector quantity. Measuring sound level by sound intensity has the following advantages compared to the sound pressure method:

- An intensity meter responds only to propagating part of a sound field and ignores the non-propagating part; for example, standing wave and reflection of surrounding walls.

- The intensity method reduces the influence of external sound sources, as long as their sound level is approximately constant, like background noise.

The sound intensity and sound pressure have a relation of $L_i = L_p - 0.2$ [4]. It means that in an ideal free field, the sound pressure and sound intensity method will yield the same numerical values for the sound power. In a real test environment, where near field effects such as background noise and reflection exist, the sound pressure method will yield a higher value if no environmental corrections are made.

9.5.3 SOUND POWER LEVEL

The sound power level is the amount of acoustic energy radiated by a source. Sound power is the cause, sound pressure is the effect. The larger the sound power that is radiated by a source, the higher the sound pressure level at a given location. In numerical terms, the sound power level is likely to be around 20 dB greater than the average sound pressure [1]. The mathematical definition of sound power level in decibel (dB) is

$$L_w = 10 \times \log_{10}\left(\frac{w}{w_o}\right) \tag{9.17}$$

where w is emitted sound power in Watts, and w_0 is reference power of 10^{-12} Watts. The sound power level of a source cannot be measured directly. The sound power level of a transformer is calculated by using the measured sound pressure levels at a known distance from the sound source, r, the measured sound pressure levels at each point are averaged, with this average sound pressure level and measurement surface, the sound power level is calculated as follows:

$$L_w \approx L_p + 10 \times \log_{10}(S) - 10 \times \log_{10} D \tag{9.18}$$

where S is the measurement surface area, m^2, when the sound pressure level measurement is conducted in shop, $S = 1.25 \times h \times l$, where h is the height of the transformer tank, m; l is the length of the contour along which measurements were made, m. When measurement is taken in free field and far enough from the transformer, S is the sphere surface area of the radius r, $S = 4\pi r^2$; r is the distance between unit and measurement point, m. D is the directivity factor which is 1 for free space, 2 for a unit located at center of a large flat surface, 4 for a unit located near the edge joint of two large flat surfaces and 8 for a unit located near the corner formed by three large flat surfaces.

When sound power level of a source is specified, it normally means the source is placed in a free unbounded space, commonly referred to as "free field". In some real situations where the source is not in free field but near a large reflection surface such as a ground surface, the sound power radiated will be 3 dB higher than free field; 6 dB higher when the source is placed at the conjunction of two surfaces, 9 dB higher when the source is placed in the corner at junction of three surfaces [4].

The sound field radiated by a source in free field is divided into three regions, the hydrodynamic near field which from sound source to much less than one wavelength, the geometric near field and the far field. In the hydrodynamic near field, the measurements of sound pressure give an inaccurate indication of sound power of the source [4]. Based on this, if the sound power based on the measurement of sound pressure the measurement contour of which is only 0.3 meters away from sound-producing surface is inaccurate because the wavelength of 120 Hz wave is about 2.9 meters, the wavelength of 240 Hz is about 1.4 meters; both are greater than 0.3 meters. Keep in mind that the major component of the transformer sound is 120 Hz. In the geometric field, enough sound pressure measurements can determine sound power accurately. Only in far field sound pressure levels decrease, at a rate of 6 dB for each doubling of the distance from source as described in Equation (9.1).

Recent environmental requirements demand accurate sound level measurement, so it may be necessary to be able to predict the sound pressure level at a distance of, say 200 meters from the substation as Example 9.4. It is essential to know the sound power level of the transformer in this substation; with sound power, the sound pressure at any distance from the transformer can be estimated. Also, contribution from more than one source can be added.

Example 9.4

The maximum allowed sound pressure at a property line is 50 dB, the property line is 200 meters away from the transformer. The transformer is 6.5 meters high and 25 meters perimeter. Here it needs to know what the sound pressure level of the transformer should be. A transformer is considered as a point source of sound in far field. Based on Equation (9.18), the sound power of the transformer is

$$L_w = L_p + 10 \times \log_{10} S_1 - 10 \times \log_{10} D = 50 + 10 \times \log_{10}\left(4\pi \times 200^2\right) - 10 \times \log_{10} 2 = 104\,dB$$

Where L_p is the sound pressure in dB at distance r from transformer, L_w is the transformer sound power, dB, $S_1 = 4\pi r^2$; D is directivity factor, due to ground existence, $D = 2$. The measurement surface $S_2 = 1.25 \times 6.5 \times 25 = 203.1$ m². Per Equation (9.18) the transformer sound pressure should be when measured per IEEE C57.12.90

$$L_p = L_w - 10 \times \log_{10}\left(S_2\right) = 104 - 10 \times \log_{10}\left(203.1\right) = 81\,dB$$

Keep in mind that the determination of the sound power of a transformer in the free field using sound pressure measurements alone requires that for any reverberant sound to be negligible, it means no reflect component in measured sound pressure.

TABLE 9.3
Test Values of No-Load Sound and Load Sound

	60 MVA at ONAN	80 MVA at ONAF	100 MVA at ONAF/ONAF
No-load sound level (dB)	65.2	67.1	70.3
Load sound level (dB)	68.1		75.4

Example 9.5

A unit has rating of 60/80/100 MVA; the test values of load sound pressure levels and no-load sound pressure levels are listed in Table 9.3. The measurement contours of no-load sound at ONAN and ONAF are 25 and 30 meters long respectively.

1. Since the measurement contour of load sound of ONAN (60 MVA) is 30 meters, while its no-load sound measurement contour is 25 meters, the load sound has to be modified to 25 meters contour.

$$L_{p,load,60MVA} = 68.1 + 10 \times lg_{10}\left(\frac{30}{25}\right) = 68.1 + 0.8 = 68.9\,dB$$

The total sound level at rated voltage and 60 MVA is

$$L_{p,total,60MVA} = 10 \times log_{10}\left(10^{0.1\times65.2} + 10^{0.1\times68.9}\right) = 70.4\,dB$$

The total sound level at rated voltage and 100 MVA is

$$L_{p,total,100MVA} = 10 \times log_{10}\left(10^{0.1\times70.3} + 10^{0.1\times75.4}\right) = 75.6\,dB$$

2. Based on tested load sound level at 100 MVA, the load sound level at 80 MVA can be estimated as

$$L_{p,load,80MVA} = 75.4 + 40 \times lg_{10}\frac{80}{100} = 75.4 - 3.9 = 71.5\,dB$$

The no-load sound level is tested at 67.1 dB. The total sound level at rated voltage and 80 MVA is

$$L_{p,total,80\,MVA} = 10 \times log_{10}\left(10^{0.1\times67.1} + 10^{0.1\times71.5}\right) = 72.8\,dB$$

REFERENCES

1. Martin J. Heathcote, *The J & P Transformer Book*, 13th Edition, Elsevier, Newnes, Amsterdam, et al., 2007.
2. Bernard Hochart, *Power Transformer Handbook*, Butterworths, London, et al., 1982.
3. R. Feinberg, *Modern Power Transformer Practice*, John Wiley & Sons, New York, 1979
4. David A Bies, et al., *Engineering Noise Control, Theory and Practice*, 4th edition, CRC Press/Taylor & Francis Group, Boca Raton, London, New York, 2009.
5. S. V. Kulkarni, et al., *Transformer Engineering Design, Technology and Diagnostics*, 2nd edition, CRC Press/Taylor & Francis Group, Boca Raton, London, New York, 2013
6. Allis-Chalmers Mfg., *Transformer Reference Book*, 1951.
7. Christoph Ploetner, et al., Judging the efficiency of low noise power transformer, *Electricity Today*, Issue 2, 2005.
8. Transformers make less noise, *Think T&D*, Winter 2008–2009.
9. Z. Valkovic, Effect of transformer core design on noise level, *J. Phys. IV France* 8m, 1998, Pr2-603–Pr2-606.
10. IEEE Std. C57.136–2000, *IEEE Guide for Sound Level Abatement and Determination for Liquid-Immersed Power Transformers and Shunts Reactors Rated Over 500 kVA*, New York, 2000.

11. Jan Anger, et al., *An Innovative Concept for Noise Control of Shunt Reactors*, CIGRE, Brugge, 2007.

12. Manfred K., *Numerical Simulation of Mechatronic Sensors and Actuators*, Springer, New York, 2004.

13. Ramsis Girgis, et al., The sound of silence, *ABB Review* 2/2008***.

14. Xose M, López-Fernández et al., *Transformers, Analysis, Design and Measurement*, CRC Press Taylor & Francis Group, Boca Raton, London, New York, 2013.

15. IEC 60076–10, *Power Transformer, Part 10: Determination of Sound Levels*, 2001–2005.

16. M. Kavasoglu, et al., *Prediction of Transformer Load Noise, Excerpt from the Proceedings of the COMSOL Conference*, 2010, Paris.

17. IEEE Std C57.12.90–2000, *IEEE Standard Test Code for Liquid-Immersed Distribution, Power, and Regulating Transformers*, New York, 2000.

10 Autotransformers

10.1 BASIC RELATIONS

The difference between the two-circuit (or two-winding) transformer and the auto-transformer is a different internal connection, as shown in Figure 10.1. In a two-winding unit, total power is transferred through magnetic circuit which is the core, while in an autotransformer, part of the total power is transferred through an electrical circuit because there is a connection between the HV and LV windings. Such interconnection gives the autotransformer different external characteristics. For an ideal two-circuit unit, the voltages, currents and number of turns have the following relations:

$$\frac{V_1}{V_2} = \frac{N_1}{N_2}; \ I_1 \cdot N_1 = I_2 \cdot N_2 \tag{10.1}$$

For an auto-connected unit, the relations are

$$V_{HV} = V_1 + V_2 = \left(\frac{N_1}{N_2} + 1\right)V_2 = \frac{N_1 + N_2}{N_2}V_{LV}$$

$$I_1 \cdot N_1 - I_2 \cdot N_2 = 0; I_{HV}\left(N_1 + N_2\right) - I_{LV}N_2 = 0$$

where V_{HV} and V_{LV} are HV and LV terminal voltages to ground respectively, I_{HV} and I_{LV} are HV and LV line currents respectively as shown in Figure 10.1b. So

$$\frac{V_{HV}}{V_{LV}} = \frac{N_1 + N_2}{N_2}; \frac{I_{HV}}{I_{LV}} = \frac{N_2}{N_1 + N_2} \tag{10.2}$$

For real transformers, when leakage impedance voltage and exciting current are so small that they can be neglected, the aforementioned relations also apply to them. The definition of co-ratio of autotransformer is

$$r = \frac{V_{HV} - V_{LV}}{V_{HV}} = \frac{I_{LV} - I_{HV}}{I_{LV}} = \frac{MVA_t}{MVA} \tag{10.3}$$

where MVA_t is power transferred through magnetic circuit, MVA is total power or nameplate power, factor r represents the relation between power transferred through magnetic circuit, MVA_t, and total power, MVA. Most common co-ratio values are in the 0.3~0.7 range.

For a two-circuit unit total power is transferred through magnetic circuit; it is $I_2 V_{LV}$ when the connection is changed to auto connection, the total power transferred is $I_{LV} V_{LV}$, a portion of which is transferred through magnetic circuit, the rest is delivered directly from the primary terminal to secondary terminal. The power ratio of two connections is

$$\frac{rating \ as \ an \ autotransformer}{rating \ as \ a \ two-circuit \ transformer} = \frac{V_{HV}}{V_{HV} - V_{LV}} \tag{10.4}$$

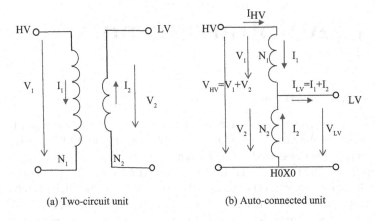

(a) Two-circuit unit   (b) Auto-connected unit

FIGURE 10.1 Operation principles of two-circuit and auto-connection units.

With same structure, material, weight and size, the autotransformer has $V_{HV}/(V_{HV} - V_{LV})$ times power of a two-circuit transformer. However, this advantage of the auto-transformer diminishes when $V_{HV} - V_{LV}$ increases, such that more power is transferred through the magnetic circuit. On the opposite, the closer the two system voltages are, the greater the gain that is achieved with autotransformer connection.

Compared with the two-circuit transformer, the autotransformer has advantages as discussed earlier, as well as disadvantages, discussed next. First with same core and coil, and same coil voltages and currents, auto-connection makes percentage impedance between HV and LV terminal, $Z_{H-L}\%$, reduced to

$$Z_{H-L}\% = Z_{2-circuit}\% \times \left(\frac{V_{HV} - V_{LV}}{V_{HV}} \right) \tag{10.5}$$

Where $Z_{2-circuit}\%$ is the impedance of two-circuit connection. Although low impedance brings better regulation, less eddy current and stray losses, it causes higher short-circuit current, therefore higher short-circuit forces. The second drawback of an auto-connected unit is that the existence of an electrical connection between high and low voltage circuits results in both circuits being affected by electric disturbances originating from one of them. Also because of this connection, overvoltage is easily transferred from one system to another galvanically and is possibly amplified, so the neutral must be effectively grounded. The grounded neutral makes single phase-to-ground short-circuit fault possible, the forces of which are sometimes higher than three-phase-to-ground case. This is the reason the short-circuit forces is the main concern in autotransformer design. The third one is that a tap changer is usually required to operate at a high voltage level [1]. Some autotransformers have delta-connected winding either for neutral stability or supplying power to another load. The impedances between HV and the delta-winding Z_{H-y}, and between LV and the delta-winding Z_{X-y}, can be calculated as

$$Z_{H-Y}\% = Z_Y\% + \left(\frac{N_1}{N_1 + N_2} \right)^2 Z_H\% + \left(\frac{N_2}{N_1 + N_2} \right)^2 Z_X\%$$

$$Z_{X-Y}\% = Z_X\% + Z_Y\% \tag{10.6}$$

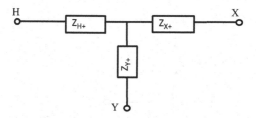

FIGURE 10.2 Equivalent T-network of positive sequence impedances.

where N_1 and N_2 are the number of turns of series and common windings as shown in Figure 10.1b, Z_H, Z_X and Z_Y are the impedances from equivalent impedance T-network as shown in Figure 10.2.

In order to reduce short-circuit current, higher impedance should be selected. Higher impedance means that the unit to be designed has a relatively small core and a high mass ratio of copper/core steel. However, increasing the mass ratio of copper/core steel too much could make the design deviate too far from a good economical design. The best way is to compromise these two aspects. In general, autotransformers have lower impedance than equivalent two-circuit units, so higher short-circuit currents as well as short-circuit forces are expected. In a large unit, a current limiting reactor (CLR) can be used in TV circuit to reduce short-circuit currents and forces.

Example 10.1

An autotransformer has base power 250 MVA, line-to-line voltages of HV and LV terminals are 345 kV and 161 kV respectively. The current in each winding and each terminal is listed in Figure 10.3.

1. The Power Transferred through the Magnetic Circuit and Directed Delivery

   From the LV side, it is $161 \times 478.1 \times \sqrt{3}/1000 = 133.3\,MVA$; from the HV side, it is calculated as $(345 - 161) \times 418.4 \times \sqrt{3}/1000 = 133.3\,MVA$; or, based on Equation (10.4), it is $(345 - 161)/345 \times 250 = 133.3\,MVA$. The power delivered directly from the HV terminal to the LV terminal is then $250 - 133.3 = 116.7\,MVA$.

2. Impedances

   The calculated impedances of a two-circuit connection at 133.3 MVA are $Z_{S-C}\% = 12.05\%$ between series and common coils, $Z_{S-T}\% = 17.04\%$ between series and tertiary coils, $Z_{C-T}\% = 3.8\%$ between common and tertiary coils. When HV and LV windings are auto connected, and the power is increased to 250 MVA, the impedance between high voltage and low voltage terminals per Equation (10.5) is

$$Z_{H-L}\% = \left(\frac{V_{HV} - V_{LV}}{V_{HV}} \right) \cdot Z_{S-C}\% = \left(\frac{345 - 161}{345} \right) \times 12.05 = 6.43\%$$

The impedances of the T-network are

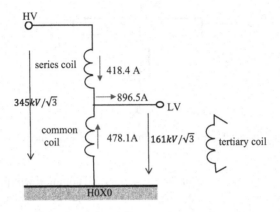

FIGURE 10.3   Autotransformer connection for Example 10.1.

$$Z_S\% = \frac{1}{2}\left(Z_{S-C} + Z_{S-T} - Z_{C-T}\right) = \frac{1}{2}\left(12.05 + 17.04 - 3.8\right) = 12.65\%$$

$$Z_C\% = \frac{1}{2}\left(Z_{S-C} + Z_{C-T} - Z_{S-T}\right) = \frac{1}{2}\left(12.05 + 3.8 - 17.04\right) = -0.6\%$$

$$Z_T\% = \frac{1}{2}\left(Z_{C-T} + Z_{S-T} - Z_{S-C}\right) = \frac{1}{2}\left(3.8 + 17.04 - 12.05\right) = 4.4\%$$

The impedance between high voltage winding and tertiary winding at 133.3 MVA per Equation (10.6) is

$$Z_{H-T}\% = 4.4 + \left(\frac{345 - 161}{345}\right)^2 \times 12.65 + \left(\frac{161}{345}\right)^2 \times \left(-0.6\right) = 7.87\%$$

It is 14.76% at 250 MVA. The impedance between low voltage winding and tertiary winding at 250 MVA is

$$Z_{L-T}\% = \frac{250}{133.3} \times Z_{C-T}\% = 7.13\%$$

Example 10.2

Regarding the unit in Example 10.1, three design options exist. In option 1, HV tap winding is placed between common and series windings as shown in Figure 10.4. The winding conductors receive high short-circuit forces and have higher eddy current loss since they are in the main leakage flux channel. From manufacturing, bringing HV tap leads needs more work because of their high BIL level. In option 2, HV tap winding is placed at the outmost, and the HV line lead is brought out through the gap between top half and bottom half of the tap winding; this needs a delicate lead insulation structure. Compared to option 1, the weight of core steel of option 2 is reduced, but the weight of copper is increased. Option 3 has the same winding disposition as option 2, but the current density in TV (tertiary) winding is increased. Thus, the weights of both core and copper are reduced. To reduce the high short-circuit forces caused by higher TV winding current density, a current limiting reactor (CLR) may be used. Even when adding the material of CLR, there is still a saving on material.

TABLE 10.1

Core steel and copper comparison

Design	Base MVA	Core Weight (kg)	No-load loss (kW)	Copper Weight (kg)	Load loss (kW)	Impedance Z (%)
Option 1	250	89686	122	30702	222	6.43
Option 2	250	82376	113	32956	224	6.29
Option 3	250	81002	111	30093	222	6.47

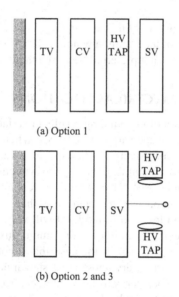

(a) Option 1

(b) Option 2 and 3

FIGURE 10.4 Winding dispositions (TV is tertiary winding, CV is common winding, SC is series winding).

With the same nameplate MVA rating, the impedance of the auto-connected unit is r^2 (r is the co-ratio of the autotransformer) times less than a two-circuit unit, however, the low impedance doesn't indicate that auto-connected unit has low leakage flux; the leakage flux relates to impedance divided by r^2, as the result, the leakage flux of the auto-connected unit could be high with small impedance. The effects of this leakage flux must be carefully considered in order to eliminate overheating problems of the tank wall, core clamping structure and outer pocket of core, due to excessive eddy loss in these parts produced by the leakage flux.

10.2 INSULATION CONSIDERATION

When there is a line-ground fault on LV, the HV system voltage will develop across the series winding only, as the common winding is shorted. This happens frequently in the life of a transformer, during lightning strikes on the transmission lines [1]. The autotransformer is particularly sensitive to lightning impulse voltage, because with the same voltage level, the series winding has fewer turns than one in HV winding of two-circuit units, while both are subjected to the same overvoltage.

If a tap winding is located in a line end, usually LV line end to regulate LV voltage only, at nominal tap position where the tap winding is overhanging and not carrying current, a transient resonance could occur, thus taking the free end above the applied voltage level. In the absence of any sophisticated computer programs, the highest voltage to earth attained by the tap winding, V_{tap}, may be estimated by the following equation:

$$V_{tap} = V_{LV} \left(\frac{N_{tap} + N_{com}}{N_{com}} \right) \tag{10.7}$$

Where N_{tap} is the number of turn of the tap winding, V_{LV} is LV line to ground voltage, N_{com} is the number of turn of the common winding. This oscillation can be limited by metal oxide varistor, but it is not preferred because maintenance in service.

10.3 TAP WINDING ELECTRICAL LOCATIONS

Different electrical location of tap winding results in variable flux voltage variation (VFVV) or constant flux voltage variation (CFVV). When tap winding locates next to the neutral end, as shown in Figure 10.5a, it creates a variable flux under constant HV voltage. Adjacent to the neutral point, the tap winding and tap changer of the lower insulation level can be used. Also, a small current tap changer can be selected due to the current in common winding being less than the LV line current. For a VFVV design, the no-load loss and sound level will vary at each tap position. The core is a little bigger to compensate for the maximum flux density, and the tertiary winding voltage varies at each tap. In the case that a constant tertiary voltage is needed, a series transformer (tertiary booster), as shown in Figure 10.5a1, is added. The variable voltage of the booster adds to or subtracts from the tertiary to maintain constant tertiary voltage under no-load conditions. The impedance of this booster should be taken into account when determining the reactance to the tertiary.

When tap winding is located between series and common windings and LV voltage is assumed constant, HV voltage is regulated, as shown in Figure 10.5b. This connection gives CFVV. The tap winding's current is the same as the series winding. The tap changer and tap winding must be insulated for the voltage level of LV line terminal plus the voltage across the tap winding. Compared with tap winding being next to neutral, higher voltage tap winding and high voltage tap changer are required.

When tap winding locates between the LV line terminal and auto point, as shown in Figure 10.5c, LV voltage is regulated by the tap winding and HV voltage is constant. This design is CFVV. The tap winding carries the full LV line current. The tap changer and tap winding must be insulated for the LV line terminal voltage level plus the voltage across the tap winding. Compared with tap winding next to neutral, higher voltage, a high current tap winding and tap changer are required. By inserting a series transformer in the LV line as shown in Figure 10.5c1, a smaller size tap changer could be used. This layout also gives flexibility on tap winding design.

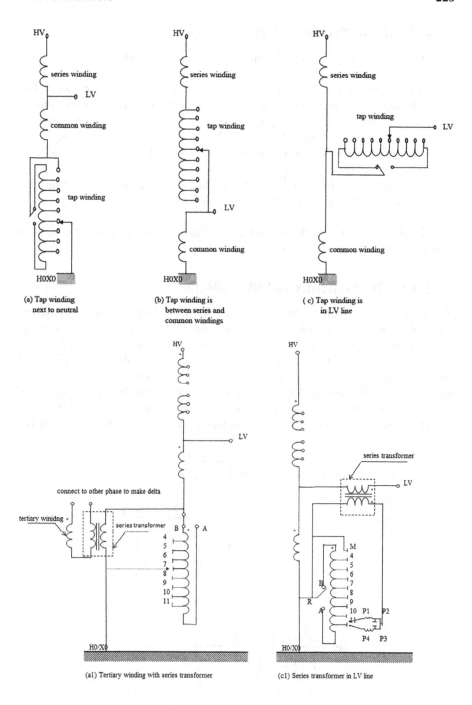

FIGURE 10.5 Tap winding electrical locations.

10.4 WINDING PHYSICAL DISPOSITION

Commonly used winding physical dispositions are listed as follows:

- *Core/tap/com/series*. The tap winding is situated outside the main leakage flux channel, so the impedance swing is usually small. Subdivision of the conductors of tap winding is not important. Tap winding has low short-circuit forces.
- *Core/common/tap/series*. The tap winding is situated in the main leakage flux channel. Subdivision of the conductor is needed. The tap winding has high short-circuit force. There is large swing of the impedance.
- *Core/common/series/tap*. The tap winding is outside the main leakage flux channel, so the tap winding has low short-circuit forces.

Tertiary winding is usually situated next to core in most designs but sometimes appears in other places like outmost.

10.5 USE OF AUXILIARY TRANSFORMERS

When LV winding has high current and its voltage regulating is required, auxiliary transformer is sometimes used. It also gives flexibility of designing tap winding. There are two common ways to use the auxiliary transformer. First, a step voltage regulator, as shown in Figure 10.6, is used to regulate the LV voltage, the tap winding in step voltage regulator offers voltage regulation, the main transformer consists just of LV and HV windings, which can be optimized independently. The second way is to use a series transformer to shunt part of the LV current away from tap winding, as shown in Figure 10.7. The tap winding is wound on the main core as LV and HV

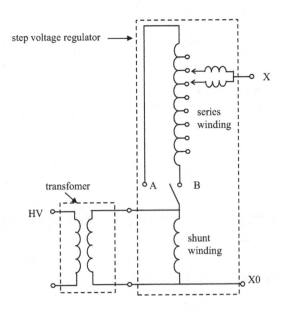

FIGURE 10.6 Connection of step voltage regulator with transformer.

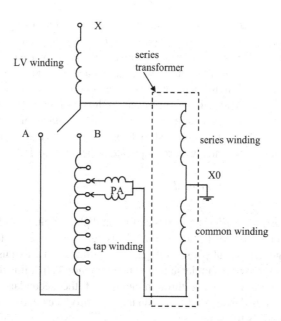

FIGURE 10.7   Tap winding connection with series transformer.

windings. This way, the tap winding has flexibility regarding number of turns per tap and amperage. Because only a portion of LV current flows into tap winding and switch, a smaller switch can be used. Sometime such a design is necessary when the current exceeds the switch limit. Both methods are applicable to both auto-connected unit and two-circuit units.

10.6 ZERO SEQUENCE IMPEDANCE AND DELTA-CONNECTED WINDING

As any type of Y–Y connected unit, an autotransformer has inherent neutral instability. The neutral instability is the result of the fact that the currents flowing in branches of an ungrounded Y connection are not independent of each other, the current entering one phase must flow out through the other two phases. Three things that could make neutral unstable are magnetizing current, line-to-ground load and third harmonic current.

- *Magnetizing current.* In order to generate the same voltage on each secondary phase, the magnetizing current in each phase should be different because each phase's magnetic circuit has different reluctance. However, the symmetrical primary voltages force the magnetizing current in each phase to be the same and vectorially sum to zero. As a result, the phase which requires less current to generate normal flux receives more current and generates more flux and voltage. This applies to single-phase and three-phase five-leg or shell-type transformer. It doesn't affect the three-phase three-leg core type unit to an appreciable extent because of its interlinked magnetic structure.

- *Phase-to-neutral load.* When a secondary winding has a single-phase line-to-neutral load, and the primary neutral is not connected to the neutral of its source, the corresponding primary current of such load has to flow through other two phases. Because other two phases have no load, the currents flowing through these primary windings become magnetizing currents which are greater than the magnetizing current in the loaded phase. As a result, the voltages of these two phases increase while the voltage of the loaded phase decreases, making the neutral shift from its geometrical center. The shifted neutral voltage depends on zero sequence impedance and is

$$\%neutral\ shift = \frac{\%I}{300} \cdot \%X_0 \qquad (10.8)$$

where % *neutral shift* is percentage of normal phase voltage, $\%I$ is single phase load in percentage of full load and $\%X_0$ is zero sequence reactance voltage in percentage of normal phase voltage. This voltage is non-sinusoidal in waveform. Another way to understand the phenomenon is that this unbalanced load makes a current flowing through neutral of the secondary side; the zero sequence impedance on the secondary side produces zero sequence voltage drop that shifts the neutral.

- When the neutral of the Y–Y connected unit is not connected to the source neutral, there is no path in the unit to supply the third-harmonic magnetizing current and its multiples, which are needed to generate sinusoidal wave flux and induced voltage; it is then supplied by line-to-ground capacitance in the supply system. If this path is also not available, the third-harmonic currents are completely suppressed because there is no path for them to flow, the magnetizing current becomes sinusoidal, resulting in non-sinusoidal wave flux and induced voltage which contain third harmonics components. These third-harmonic voltages cannot be summed to zero but shift the neutral from its geometrical center. Three-phase three-leg core form units are less susceptible to this neutral shift, because of its inherent virtue of lower zero sequence impedance.

It can be seen from the foregoing discussion that to reduce neutral shift, the zero sequence impedance between line and neutral is needed to be small, which can be achieved by the following methods:

1. Directly connect the neutral of the transformer primary side to its source neutral, make each phase perform independently, in such way, each phase-to-neutral voltages always remain balanced regardless of the degree of unbalanced phase-to-neutral loads. However, transmission cables have significantly higher zero sequence impedance per mile than positive sequence impedance, making connecting two neutrals through low impedance difficult [2].

2. Add a delta-connected winding to stabilize the neutral. The zero sequence currents either by third harmonics of the magnetizing current or single-phase line-to-neutral load flow in the closed delta winding, reducing zero

sequence impedance significantly. The neutral shift is improved but not fully balanced because the stabilizing winding itself has zero sequence impedance. The shifted neutral voltage can be known by Equation (10.8). As can be seen, the smaller the zero sequence impedance is, the more stable the neutral is.

In additional to reduced neutral shift caused by single-phase line-to-neutral load, the delta-connected winding can consume third harmonics voltage by circulating current in its loop, suppressing third harmonic voltage and current appearing in line to interfere with a telephone communication line. However, the magnetizing current is usually less than 1% by using cold-rolled grain-oriented core steel without a PA (preventive autotransformer) and/or a series transformer, less than 1.3% with them, the delta winding as stabilizing winding may be eliminated if the interference suppression is the only reason. Further, the telephone interference problem is not as severe as it was years ago, as open telephone circuits with ground return have been replaced by open circuit with metallic return or by cable.

Some autotransformers and Y–Y connected two-circuit transformers have delta-connected winding. This plays two roles: stabilizing the neutral, and/or supplying power to the external load. When none or only one corner of the delta winding is brought out through one or two bushings and is earthed externally, it doesn't supply power to external load; its function is to stabilize the neutral when there is single-phase line-to-neutral load, or when there is single phase-to-ground fault, as well as consume the third harmonic voltage by circulating the third harmonic magnetizing current in the winding, this delta winding is called stabilizing winding (SW). If the single-phase line-to-neutral loading is continuous, the rating of the SW should be such that it performs thermal duty as requested. Line-to-ground fault on primary or secondary sides of the unit causes a high circulating current flowing through the SW. It has to have enough mechanical strength to withstand the forces and thermal capability to absorb the heat resulting from such short-circuit current, for example, winding copper conductor temperature during 2 seconds short-circuit should be lower than 250°C. The stabilizing winding is usually called buried tertiary winding.

When three corners of the delta winding are brought out through three bushings, besides its inherent ability to stabilize the neutral of wye-connected windings, it also supplies power to the external load such as reactive compensation equipment or other equipment in substation. This winding is usually called load tertiary or tertiary windings (TW). From a short-circuit point of view, besides high circulating current resulting from HV or LV single phase-to-ground fault, the tertiary winding behaves as a stabilizing winding; the tertiary winding has potential for its own three phases-to-ground fault. The tertiary winding has to have enough mechanical strengths and thermal capability to withstand both types of faults. Also, by having tertiary winding, the rating of either series winding or common winding has to be increased based upon whether the unit is for step-up or step-down operation. When the unit is designed for bidirectional power flow, step-up and step-down, both series and common windings have to have the ratings to supply the power to the tertiary winding and meet thermal or cooling requirements.

3. Use a three-phase three-leg core. Zero sequence current produces zero sequence flux; this flux has to leave from one yoke, going through the space outside the core and back into another yoke. High reluctance on zero sequence flux path in the space needs high magnetizing current to drive the flux; as a result, it produces a relatively low zero sequence magnetizing impedance, it makes the unit able to supply considerable single-phase and line-to-neutral load without very severe neutral shift [1]. Three-leg core reduces the zero sequence impedance in a fashion normally not as effective as a stabilizing winding does which may have single-phase line-to-neutral load up to rated current [3], but it does provide some degree of stability at a lower cost.

Three-phase five-leg core, three-phase bank of single phase units without delta winding, and shell form core type units give zero sequence flux a low reluctance path: side legs, very low magnetizing current is then needed, as the result, zero sequence impedance being magnetizing impedance is very high as shown in Figure 10.8a. Any single-phase line-to-neutral load can shift the neutral easily. When a stabilizing winding is added, a path for zero sequence current is offered, and the T-network of zero sequence impedance is as shown in Figure 10.8. As can be seen, the effect of core magnetizing impedance is reduced by the stabilizing winding impedance which parallels with it. In terms of zero sequence impedance value, for a five-leg core with stabilizing winding,

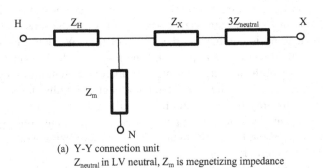

(a) Y-Y connection unit
$Z_{neutral}$ in LV neutral, Z_m is megnetizing impedance

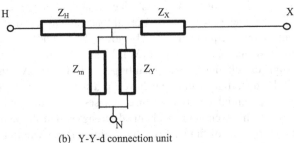

(b) Y-Y-d connection unit

FIGURE 10.8 Zero sequence impedance T-networks.

it is nearly equal to its positive sequence impedance; for a three-leg core with stabilizing winding, its zero sequence impedance is about 85–95% of its positive sequence impedance; with stabilizing winding, the types of core structure have no significant effect on zero sequence impedance to neutral.

4. Y-Zigzag connection. By the virtue of inherent low zero sequence impedance of zigzag winding, single-phase line-to-neutral load is allowed. Keep in mind that the single-phase line-to-neutral load in such a connection can shift the neutral, too, but not to a severe degree.

5. Other neutral stabilizing methods, such as a zigzag connected grounding transformer.

Compared to method 3, method 2 provides freedom to select the value of zero sequence impedance as desired and get it more accurately, but at higher cost.

10.6.1 Stabilizing Winding Rating

The maximum load of stabilizing winding (SW) under continuous single-phase line-to-neutral loading could reach 1/3 or 33% of the secondary winding load that is transferred through the magnetic circuit in extreme cases. In most cases SW load is less than that; for example, for three-phase three-leg core type units, SW load is less than 33% of the secondary winding load since the core form can reduce the SW's ampere-turns contribution, therefore, selection of 33% of secondary winding rating as SW rating in these cases could cost potentially and unnecessary. As old rule of thumb, compared to unit without SW, SW of 35% rating costs is about 110%, tertiary winding of 35% rating cost is about 120%. The Y–Y connection transformer doesn't usually need SW if its zero sequence and third harmonic characteristics are compatible with the system into which it is to be connected, or if without SW, the transformer is still capable of performing reliably under expected transient and emergency conditions.

Several methods are used to dimension SW. One of them, defined as Method B in IEEE C57.158, is discussed in detail. The circulating current in SW when there is a continuous single-phase line-to-neutral load on LV winding depends on the load current in LV and zero sequence impedances; this circulating current is I_{Y0}, shown in Figure 10.9. Similar to single phase-to-ground fault, when the load has rated current I_a, the positive, negative and zero sequence currents has the relation as

$$I_{a+} = I_{a-} = I_{a0} = \frac{I_a}{3}; \; I_a = I_{a+} + I_{a-} + I_{a0} \tag{10.9}$$

$$I_{a0} = \frac{100}{Z_+ + Z_- + Z_0 + 300} \times I_a \tag{10.10}$$

Where Z_+, Z_- and Z_0 are positive, negative and zero sequence impedances in percentage, $Z_0 = Z_{X0} + Z_{Y0}//(Z_{H0} + Z_{HSYS0})$, The current flow in SW is

$$I_{Y0} = I_{a0} \times \frac{Z_{H0} + Z_{HSYS0}}{Z_{H0} + Z_{HSYS0} + Z_{Y0}} \tag{10.11}$$

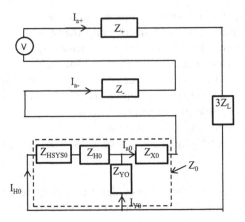

FIGURE 10.9   Impedance network of single-phase line-to-neutral load.

Where Z_{HO}, Z_{XO} and Z_{YO} are components of transformer zero sequence impedance T-network in percentage. HV system zero sequence impedance in percentage, Z_{HSYS0}, is connected to Z_{HO}, as shown in Figure 10.9, when Z_{HSYS0} goes to infinite, I_{YO} = I_{a0}, which is about 33% of LV load. However, the ratio of system zero sequence to positive sequence impedances is in the range 1~2 per reference [4], meaning that the system zero sequence impedance is a finite value, $I_{YO} < I_{a0}$. For a three-leg core, lower zero sequence magnetizing impedance is offered, making the current of SW even less. The SW rating should be the product of its rated voltage and circulating current generated in it by the single-phase load at the LV or HV side. Of course the winding shall satisfy the requirements of thermal and short-circuit strengths.

Example 10.3

An auto unit has a base rating of 240 MVA, HV voltage is 230 kV and LV voltage is 138 kV. The positive sequence impedances are, $Z_{HX, +}\% = 6.8\%$, $Z_{HY,+}\% = 27.93\%$, $Z_{XY,+}\% = 18.44\%$. The SW voltage is 13.2 kV and its rating is specified as 33.6 MVA, 35% of common winding, calculated as the following:

$$MVA_{TV} = \frac{230-138}{230} \times 240 \times 35\% = 33.6\,(MVA)$$

The SW capacity for LV single-phase rated load is calculated as following, the impedances of the positive sequence T-network:

$$Z_{H,+}\% = (6.8+27.93-18.44)/2 = 8.15\%$$

$$Z_{X,+}\% = (6.8+18.44-27.93)/2 = -1.35\%$$

$$Z_{Y,+}\% = (27.93+18.44-6.8)/2 = 19.79\%$$

Assuming zero sequence impedance is 85% of its positive sequence impedance, each component of zero sequence impedance T-network is then $Z_{HO}\% = 8.15 \% \times 0.85 = 6.93 \%$; $Z_{XO}\% = -1.35 \% \times 0.85 = -1.15 \%$; $Z_{YO}\% = 19.79 \% \times 0.85 = 16.82\%$.

The HV system has 33,530 MVA short-circuit apparent power, the HV system positive sequence impedance is 240/33,530 = 0.72%, assume the zero sequence impedance is same as positive sequence impedance, Z_{HSYS0}=0.72%, then

$$Z_0\% = (-1.15) + \frac{(6.93 + 0.72) \times 16.82}{(6.93 + 0.72) + 16.82} = 4.1\%$$

The circulating current in SW when a single-phase line-to-neutral load at LV side is

$$I_{a0} = \frac{100}{6.8 + 6.8 + 4.1 + 300} \times I_a = 0.315 I_a$$

$$I_{Y0} = 0.315 I_a \times \frac{6.93 + 0.72}{6.93 + 0.72 + 16.82} = 0.098 I_a = 0.098 \times 1004 = 98.4\,A$$

Where 1004 is LV current, I_{Y0} is SW current equivalent to LV winding and the real current in SW is

$$I_{SW} = 98.4 \times (138 / \sqrt{3}) / 13.2 = 594\,A$$

The SW rating is 594 × 13.2 × 3/1000=23.52 MVA. As can be seen, 35% rule or 33.6 MVA is higher than needed for LV single-phase line-to-neutral load. 33.6 MVA stabilizing winding is not only too big for real situation, but also it makes common and series windings bigger. It also indicates that accurate design values of zero sequence impedance help on deciding accurate SW rating.

Regarding stabilizing winding power rating, one-third of secondary winding as a rule of thumb is a rough estimate. A truly satisfactory value of the stabilizing winding can be derived only with full knowledge of the impedances between windings, system characteristics and details of the grounding arrangement. It is suggested to design SW considering a full single-phase line-to-neutral load in the largest secondary winding. The SW has to satisfy mechanical and thermal stresses caused by current circulating during line-to-ground fault at either LV or HV sides. If a continuous single-phase load between secondary line terminal and neutral is expected, the SW shall be designed to meet the thermal duty for such loading, in addition to the short-circuit withstand capabilities mentioned earlier. Placing a current limiting reactor (CLR) in a series with stabilizing winding as shown in Figure 10.10 can reduce fault currents; as a result, short-circuit forces are reduced. Large units having small impedances often use CLR to reduce mechanical stresses cause by single phase-to-ground faults. CLR has different characteristics from the main transformer as it has a separate flux path. It does, however, have an influence on the impulse voltage between SW and ground, and higher clearance may be required.

Any decision to omit stabilizing winding from Y–Y connected transmission transformer or autotransformer would only be taken following careful consideration of anticipated third-harmonic current in the neutral, the third-harmonic voltage at the secondary terminals and the resultant zero-sequence impedance to ensure that all of these were within the prescribed values for the particular installation.

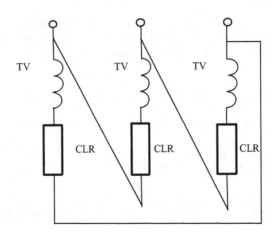

(a) CLR inside delta connected windings

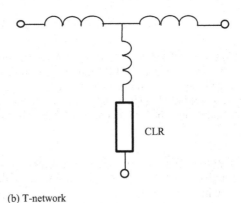

(b) T-network

FIGURE 10.10 CLR winding connection to delta winding.

When a single phase load is connected from line to line, the neutral does not shift because no current flowing in neutral and the allowable loading is referred to rated current of the respective windings.

10.6.2 LOADED TERTIARY WINDING

When delta-connected tertiary winding carries a load at the same time as LV or HV windings carry loads, this winding is usually called loaded tertiary. With loaded tertiary, the current in common winding in step-down operation, or the current in series winding in step-up operation relates to the tertiary rating, voltage as well as load characters. The following example studies three loading cases which have different power factors. In order to simplify the calculation, the exciting current and voltage drop by impedance are not taken into account.

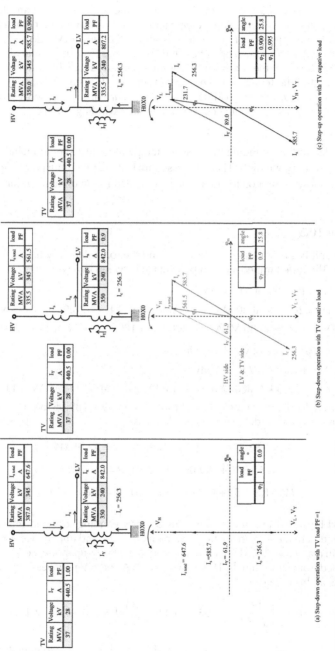

FIGURE 10.11 Loading current with different TV loads.

Example 10.4

An autotransformer, 350 MVA, 345 kV of HV voltage and 240 kV of LV voltage, has loaded tertiary of 37 MVA, 28 kV voltage. Three operation cases are studied. In case 1, a step-down operation, power factors of LV load and tertiary load are PF = 1, as shown in Figure 10.11a. HV load is simply the sum of LV load and tertiary load. In case 2, a step-down operation, the LV load has PF = 0.9, tertiary load has PF = 0, the current in the series winding is the vectorial sum of the common winding current and the tertiary winding current. In case 3, a step-up operation, HV load has PF = 0.9, tertiary load has PF = 0, the current in common winding is vectorial sum of series winding current and tertiary winding current.

With loaded tertiary winding, the loss for temperature test should include the load loss from the tertiary winding. The load losses are tested in winding pairs. The following example shows a method to obtain the loss needed for an oil temperature rise test.

Example 10.5

An auto unit has base rating of 115 MVA, and the loaded tertiary has 46 kV and 80 MVA. The tested impedances in percentage between two windings are

HV-LV = 19.63% at 115 MVA

HV-TV = 20.14% at 80 MVA = (115/80) × 20.14% = 28.95% at 115 MVA

LV-TV = 5.18 % at 80 MVA = (115/80) × 5.18% = 7.45% at 115 MVA

The tested load losses of winding pairs are

HV-LV = 83.376 kW at 115 MVA

HV-TV = 102.4 kW at 80 MVA = (115/80)² × 102.4 = 211.59 kW at 115 MVA

LV-TV = 91.49 kW at 80 MVA = (115/80)² × 91.49 = 189.06 kW at 115 MVA
The loss of each winding can be calculated as

$$HV = (83.376 + 211.59 - 189.06)/2 = 52.95\,kW$$

$$LV = (83.376 + 189.06 - 211.59)/2 = 30.42\,kW$$

$$TV = (211.59 + 189.06 - 83.376)/2 = 158.64\,kW$$

It should be noted that the calculations of each winding loss and impedance conducted here are a mathematical treatment; the loss and impedance exist only in pairs in reality. The load loss has its highest value at step-up mode, i.e., 115 MVA is supplied from LV terminals, TV receives 80 MVA and HV receives 35 MVA. The load loss in this case is

$$Load\ Loss = \left(\frac{35}{115}\right)^2 \times 52.95 + 30.42 + \left(\frac{80}{115}\right)^2 \times 158.64 = 112.1\,kW$$

The tested no-load loss is 35.58 kW. So the total loss for oil temperature rise test is

$$Total\ loss = 112.1 + 35.58 = 147.68\,kW$$

There are two ways to conduct the test. One is supplying power to LV terminals, HV terminals are short-circuited, TV terminals are open. To avoid overloading the windings, load loss of HV-LV is used, the total loss is 83.376 + 35.58 ≈ 119 kW (about 80.6% of total loss required). The heat run test is conducted with 119 kW. The tested oil temperatures are then corrected to 147.68 kW.

Another way is supplying power to LV terminals. Both HV and TV terminals are short-circuited. Under this situation, the load distribution between HV and TV is determined by their impedance ratio. The impedance of each winding is

$$HV = (19.63 + 28.95 - 7.45)/2 = 20.56\%$$

$$LV = (19.63 + 7.45 - 28.95)/2 = -0.93\%$$

$$TV = (28.95 + 7.45 - 19.63)/2 = 8.39\%$$

The load distribution is

$$HV = \frac{8.39}{20.56 + 8.39} \times 115 = 33.33\,MVA; TV = \frac{20.56}{20.56 + 8.39} \times 115 = 81.67\,MVA$$

Compared with their request loads,

$$HV = \frac{33.33}{35} = 95.23\%; TV = \frac{81.67}{80} = 102.08\%$$

HV is underloaded, and TV is overloaded. When the exact load distribution is request, the load distribution should be

$$HV = \frac{35}{115} = 30.43\%; TV = \frac{80}{115} = 69.57\%$$

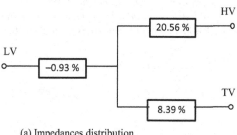

(a) Impedances distribution

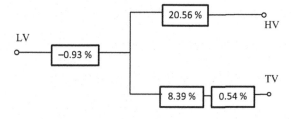

(b) Impedances distribution with external impedance

FIGURE 10.12 Impedance distributions of a three-winding unit.

In order to get such a load distribution, the impedance ratio of HV and TV should be HV/TV = 69.57/30.43 = 2.29, while the real impedance ratio of HV and TV is 20.56/8.39 = 2.45. The impedance in TV circuit has to be increased to TV = 20.56/2.29 = 8.98%. An external reactor with impedance value of 8.98% − 8.39% = 0.59% is needed. At 115 MVA, TV winding has 833.3 A current. The voltage drop at the reactor is 0.59% × 46000 = 271.4 V. The reactor impedance value in Ohms is $R_{external}(\Omega) = 271.4/833.3 = 0.326\Omega$. The test lab has 0.3Ω reactor available, and its impedance in percentage is

$$R_{external}\left(\%\right) = 0.3 \times 833.3 / 46000 = 0.54\%$$

The impedance of the TV winding circuit is now 8.39 + 0.54=8.93%. The load distribution on HV and TV is now

$$HV = \frac{8.93}{20.56 + 8.93} \times 115 = 34.8\,MVA; TV = \frac{20.56}{20.56 + 8.93} \times 115 = 80.2\,MVA$$

Compared with their requested loads

$$HV = \frac{34.8}{2435} = 99.4\%; TV = 80.2 / 80 = 100.3\%$$

This distribution is closed to the request.

REFERENCES

1. Martin J. Heathcote, *The J & P Transformer Book*, 13th edition, Elsevier, Newnes, Amsterdam, et al, 2007.
2. IEEE Std C57.158-2017, *IEEE Guide for the Application of Tertiary and Stabilizing Windings in Power Transformer*, New York, 2017.
3. IEC 60076-8, *Power Transformer – Application Guide*, 1997–10.
4. IEEE Std C57.12.00-2015;,*IEEE Standard for General Requirements for Liquid-immersed Distribution, Power, and Regulating Transformers*, 2016.

11 Testing

The purpose of testing prior to ship is to ensure a transformer meets specified performances such as insulation levels, impedance, no-load loss, load loss, temperature rise and sound levels, and that it has built-in qualities. Here, a brief discussion on testing is given.

11.1 PRELIMINARY TESTS

Preliminary tests include ratio, polarity, winding DC resistance, power factor and insulation resistance (megger) tests. These are conducted before power tests.

11.1.1 RATIO AND POLARITY

This test is to check transformation ratio (high voltage per phase/low voltage per phase) and phase relation; the information on the nameplate serves as references, the TTR (transformer turn ratio) bridge is usually used, its high and low voltage test leads are connected to high and low voltage terminals of the transformer respectively, ensuring phase relation reflect the vector relation on nameplate, the test should be conducted at all taps. It is suggested to apply higher voltage in order to get more accurate test results. A unit passes the ratio test if the tested value is within ±0.5% of the designed value and phase deviation is less than 1% [1]. If ratio error is higher than 0.5%, check related standards, specifications and design to see if it is acceptable.

TTR usually has 100 Volts as maximum voltage. When a ratio test needs higher voltage, say, 10 KV as an example, the so-called voltage method is applicable; it reads both primary-side and secondary-side voltages simultaneously, then calculates the ratio based on the readings. The voltage method is often used for failure investigation to find out voltage-related issues. The polarity test can be done by voltmeters, connecting a LV terminal to a HV terminal and measuring the potential differences that appear between the open terminals and from which the phasor diagram can be derived. The polarity test can also be done using DC current.

11.1.2 WINDING DC RESISTANCE

The volt/amp method is most often used to test winding DC resistance; another method is to use Kelvin (Thomson) bridge. By Ohm's law, DC resistance of a circuit is that the DC potential drop between the circuit terminals is divided by the DC current flow through the circuit. Winding resistance values are in a range from 0.0002 to 200 ohms depending on voltage and power rating of the unit. To test a low resistance with high accuracy, the four-electrode method is used; two electrodes are for current flowing, and other two are for voltage measurement which are connected directly to bushing terminals, in such way that error caused by the resistance of test leads is eliminated. 40 to 100 amps may be applied to get an accurate voltage reading, while the test current is also limited to 15% of winding rated current. Because winding presents not only a resistance but also a large inductance, when a voltage is applied across its two terminals, the current needs time to get to saturation level. The readings of voltage, current or resistance are taken when they are stable. To protect the voltmeter from damage due to off-scale deflections, the voltmeter should be disconnected from the circuit before switching the DC current on or off.

Winding resistance is temperature dependent, and the winding temperature is assumed to be average oil temperature under such conditions. The oil temperature has to be stabilized prior to the test, meaning that the top oil temperature does not vary more than 2°C in a one-hour period, and the difference in temperature between top and bottom does not exceed 5°C [1]. The winding shall have no excitation and no current for a minimum of three hours.

During the temperature rise test, hot resistance measurements are taken to get winding temperatures which also relate to the cold resistances and its temperatures, so it is important to get accurate cold resistances and its temperatures, otherwise, any error from cold resistance measurements will be transferred into the winding temperature rise test results. The same equipment and same current value are required for testing both cold and hot resistances in order to reduce possible measurement error.

Lead and contacts such as lead join, bushing connection, tap changer, delta/wye converter and multi-voltage selector all contribute to the resistance measured from bushing terminals. Unequal results are always found between phases and need to be judged; the difference between phases is usually within 1% after leads compensation. A way to see the difference is to put the results in curves; an example is in Figure 11.1.

Directed touching or so-called wire-short between winding conductor wires could happen in manufacturing, the resistance test method can be used to locate the wire-short in a winding. Assume that the length of the wire is L, resistance is R_0, location

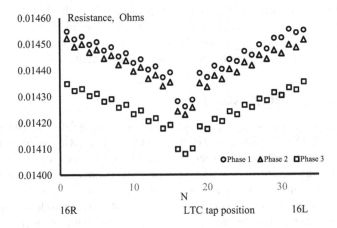

FIGURE 11.1 Three-phase winding resistances vs. LTC tap positions.

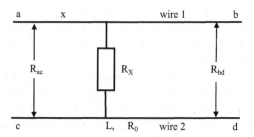

FIGURE 11.2 Locating wire short by resistance test.

of wire-short is x, and contact resistance R_x as shown in Figure 11.2. Two tests are required, test resistance R_{ac} between a and c, and test resistance R_{bd} between b and d, which are:

$$R_{ac} = 2R_0\frac{x}{L} + R_X; R_{bd} = 2R_0\frac{L-x}{L} + R_X \qquad (11.1)$$

The short location x is

$$x = \frac{L}{4R_0}\left(R_{ac} - R_{bd} + 2R_0\right); R_x = \frac{1}{2}\left(R_{ac} + R_{bd} - 2R_0\right) \qquad (11.2)$$

11.1.3 POWER FACTOR AND CAPACITANCE

It is also called dielectric loss or dissipation factor test. A capacitance bridge can be used for direct measurement [2]. Power factor and capacitance can also be measured based on tested dielectric loss, voltage and current; in this test, short bushings together in each winding group (HV, LV, TV) respectively. Figure 11.3 shows a typical sketch of a two-winding model. C_{HL}, C_L and C_H are capacitances between HV and LV, LV to ground, and HV to ground respectively. The test is conducted as follows:

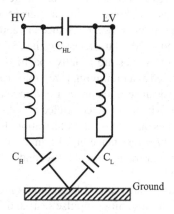

FIGURE 11.3 Two-winding transformer PF test model.

first, energize HV and ground LV to get C_{HL} // C_H; second, energize HV and guard on LV to get C_H; third, UST (unground system test) to get C_{HL}. Repeat the second test but energize LV and guard on HV to get C_L. Power factor (PF) of each capacitance is also measured in the tests [3].

Test voltage normally is 10 KV. When winding's rated voltage to ground is less than 10 KV such as LV, winding's rated voltage may be used. PF is temperature dependent; for comparison purposes, PF tested results are often corrected to 20°C, correction factor K_1 (=$1.55 \times e^{-0.022054 \times T}$) is used for test temperature T between 5°C and 50°C. For example, PF is tested 0.54% at 31°C, at 20°C, $K_1 = 1.55 \times e^{-0.022054 \times 31}$ = 0.782377, then PF at 20°C is 0.782377 × 0.54% = 0.42%.

Good insulation condition in a transformer has PF of less than 0.5%. If PF is high, the test should be repeated to investigate the cause, such as moisture or dirt in oil, or wet solid insulation. However, PF varies case by case, and careful judging is needed.

11.1.4 Insulation Resistance

It is also called the Megger test because the readings are in Mega Ohms. It tests insulation resistances of windings, cores (main, preventive autotransformer (PA), series transformer) and core clamping frame if any. A DC voltage is applied, then the resistance reading is taken. For windings, the resistances are tested at voltage up to 5000 V, for the core, up to 1000 V. The test should be conducted with the negative lead from meter connected to the test object and the positive lead connected to the ground due to water molecules in the cellulosic insulation are polar positive. Insulation resistances values relate to the voltage applying time; the reading is often taken at 60 seconds after DC voltage is applied, and the temperature is recorded.

For a two-winding unit, the measurement connections are HV-LVG, LV-HVG and HV + LV-G; here G is ground, and HV and LV are indicated in Figure 11.3. When using a guard lead in the meter, the test is carried on HV-G guard on LV, LV-G guard on HV. This is a three-electrode measuring system, using the guard lead to the terminal as required. For a three-winding unit having tertiary winding (TV), the test connections are HV-LVTVG, HV-G, HV-LVTV, LV-HVTVG, LV-G, LV-TVG, TV-LVHVG and TV-G. Terminals that are not shown in connections are on guard. Use the factor K_2 to correct insulation resistance test result at temperature T to 20°C; here, $K_2 = 0.251 \times e^{0.0689T}$. For example, tested resistance is 1250 MΩ at 24°C, $K_2 = 0.251 \times e^{0.0689 \times 24} = 1.31$, corrected to 20°C, the resistance is 1.31 × 1250=1638 MΩ.

The minimum acceptable values are usually specified by customers; otherwise, 1000 MΩ at 20°C is suggested as a minimum. Due to absorption currents of insulation material, the insulation resistance is inconstant and increases with time during the test, the phenomenon of which is called dielectric absorption. Two resistances measured at different times are used in Polarization Index (*PI*) and Dielectric Absorption Ratio (*DAR*), *PI* = resistance at 10 minutes/resistance at 1 minute, and *DAR* = resistance at 60 seconds/resistance at 30 seconds. When insulation is in good condition, *PI* ≈ 2, and *DAR* ≈ 1.25.

Core resistance is usually tested after a large unit is moved from one location to another, before and after the lightning impulse test, and is loaded on truck before shipping. If the dielectric absorption phenomenon is not observed during the test, the core ground connection may be opened; an investigation is needed.

11.2 NO-LOAD LOSSES AND EXCITATION CURRENT

No-load loss and exciting current are the loss and current consumed by the transformer, which is energized to rated voltage but without any load. No-load loss contains losses from the core and insulation materials, and I^2R loss of primary winding caused by the exciting current. Test set-up is the same as operation connection; test voltage is applied to either LV, TV or HV bushings while leaving the rest of bushings open. It is common practice to energize low voltage bushing for convenience. In some special cases, only a portion of a winding is energized; this portion shall not be less than 25% of the winding. The test is conducted with rated voltage; its frequency should be within ±0.5% of rated frequency of the test unit. When rated frequency cannot be achieved, conversion factors have to be used to convert test values from test frequency to the rated one per agreements between maker and user. For protection purposes, before the test all terminals of internal current transformers (CT) should be shortened and grounded, leaving the internal potential transformer (PT) secondary open and connecting one terminal of each secondary to ground. For a three-phase unit, a three phase four wires test system, as shown in Figure 11.4, is preferred.

The test voltage should be sinusoidal wave. In the test average absolute voltmeter method is used, and the test voltage is adjusted until the average of three voltmeter readings is the desired voltage level. When tested voltage deviates from the requested value greater than 0.1% due to equipment limits, the actual loss at requested voltage is achieved by interpolation. The voltage waveform may be distorted, and a waveform correction shall be applied to the test results as the following:

$$W_1 = W_t / \left(p_1 + k \cdot p_2 \right) \qquad (11.3)$$

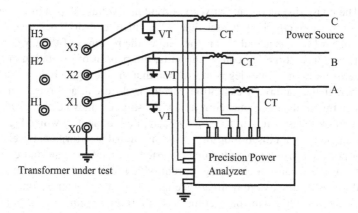

FIGURE 11.4 No-load loss and exciting current connection diagram by the three-meter method.

where W_1 is no-load loss corrected for waveform at temperature t, p_1 is per unit hysteresis loss, p_2 is per unit eddy current loss, W_t is measured no-load loss at temperature t, $k = (V_r/V_a)^2$, here V_r is the test voltage measured by rms voltmeter, V_a is the test voltage measured by average-voltage voltmeter. If waveform distortion causes the correction to be greater than 5%, the test voltage waveform needs to be improved.

Transformers may contain residual flux in its core. De-magnetization may be done before any reading is recorded, that is, the voltage is increased gradually to 110% or 115% and held for a few minutes, then is reduced gradually down to minimum before being cut off. The voltage is increased again to 110% or 115%, and held for a short time; the readings are then taken starting at 110%, 105%, 100% and 90% of rated voltage. When no-load loss curve is required, more readings are needed. The reference temperature of no-load loss is usually 20°C; the test conducted at temperature t is corrected to reference temperature by the following equation:

$$W = W_1 \times \left(1 + \left(t - t_r\right) \cdot k_t\right) \tag{11.4}$$

Where t_r is the reference temperature, W is the no-load loss corrected to the reference temperature, k_t is an empirically derived per-unit change in no-load loss per degree Celsius, and 0.00065 is used when the actual value is not available. Tested no-load loss shall not exceed guaranteed loss by more than 10% [1].

The excitation current is often expressed in percentage of rated line current, base KVA rating is used to determine the current, rms value of the current is recorded simultaneously with no-load losses. For a three-phase unit, average of three of line currents are used to determine the excitation current. For small power transformers the excitation current is about 1% to 5% of rated current at rate voltage; for large power transformers, it is in the range of 0.1% to 0.3%.

When using three wattmeters to read no-load loss, sometimes reading of one wattmeter is negative; it may happen on large unit, or when test voltage is high. The magnetic asymmetry of the core causes asymmetrical excitation currents in each phase. The phase displacement between current and voltage depending on the flux density in the core could be greater than 90°, which leads to a negative power in one wattmeter [4]. The actual loss is the sum of the readings of three wattmeters.

For three-phase three-leg core form units, the pattern of excitation currents is that the current in the center leg is lower than that in side legs which are similar to each other. If the pattern of excitation current is different, and one or more legs have much higher than expected current, an investigation may be needed to find the cause. A single-phase excitation test may be applied to identify which leg has possible issue, such as a closed loop in a winding, unbalance turns between parallel winding portions, winding conductor is shorted, a metal loop around core leg or yoke. All of these cause a circuiting current which leads to higher excitation current and losses. Burrs at core lamination edges also cause higher excitation current and losses, especially when volts per turn is high. If high excitation current is found in one or two phases, it is more likely to be winding related. If the losses are much higher than designed, all three phase windings may be involved. Core loss run at 110% voltage for 24 hours and analysis of gases in oil afterwards may give more information for judgment. Conducting the test with the tank cover removed may help to see the oil heat wave and locate the trouble. If core and coil assemble is taken out of tank, core loss is tested in air with lower voltage; a thermal scan may help to locate the hot location due to circulation current, and a hot resistance and cold resistance tests may also help to locate trouble in winding. Using a "C coil sensor" is another way to locate trouble.

Single-phase excitation current can be measured at lower voltage, a few hundred volts to 10 kV, either on LV or HV bushings. Such a test is sometimes conducted on the service site to find possible failures in winding insulation, core displacement and problems in tap changers after transport of the transformer. This is one of many auxiliary judgment methods to assess the condition of transformers. The pattern of the excitation currents is compared to the ones tested in shop.

11.3 LIGHTNING IMPULSE AND SWITCHING IMPULSE

The purpose of the impulse tests is to ensure that insulation can withstand transient overvoltages caused by atmospheric lightning and switching operations in service.

11.3.1 LIGHTNING IMPULSE WAVESHAPE AND SWEEP TIMES

Standard waveshape of lightning impulse full wave is defined as front time $t_1 = 1.2$ µs from zero to crest, tail time $t_2 = 50$ µs from zero to time of half of the crest [1,5]; the tolerances are ±3% for voltage crest value, ±30% for t_1, and ± 20% for t_2. In the test, t_1 is 1.67 times the time span between 30% and 90% of the crest, t_2 is the time span from the intersection of straight line through 30% and 90% of the crest and time axis, to the time of 50% crest. When oscillations exist in the front of the wave, the points of 30% and 90% shall be determined from the average smooth wave sketched through the oscillations.

Transformer is also tested by chopped wave. Two types of chopped waves are used, chopped at front which is the front of wave (FOW) test, and chopped at tail which is the chopped wave (CW) test. The FOW test is conducted only per specified, while the CW test is conducted with the full wave (FW, lightning impulse) test. CW's front is same as FW's but 10% higher in the crest value; chopping at tail occurs at 2 or 3 to 6 µs from start. The overshoot of the amplitude after first zero crossing should not be more than 30% of the crest. The dropping time of the chopping from start flashing to go over zero may be limited to a range. The chopping gap is set up to close to bushing terminal, or directly set up at the bushing terminal may help.

Preferred test sweep times for voltage oscillograms are 50~100 µs for FW, 5~10 µs for CW, 2~5 µs for FOW. Preferred test sweep times for the current oscillograms are 100~600 µs for FW, 25~100 µs for CW, 10~25 µs for FOW. Test on oil fill transformers are performed by negative polarity wave to reduce the risk of external flashover.

11.3.2 TEST SET-UP AND PROCEDURE

The impulse test shall be applied to each bushing and one at a time. Non-impulsed terminals shall be grounded directly or through a resistance no more than 450 Ω. When a non-impulsed terminal is grounded through resistor, the voltage on that terminal shall not exceed 80% of its FW level (BIL). The voltage should be confirmed at reduced full wave (RFW) level before staring the test.

All internal CTs' secondary terminals shall be shorted and grounded; if there are PTs inside the tank, their secondaries shall also be shorted and grounded. The tap changer shall be at position to make the winding under the test have minimum effective turns, or per standards or specification. The current shunt in the test circuit, as shown Figure 11.5, is used to record the current wave for comparison between RFW and FW, the current shunt is connected between neutral terminal of wye-connected windings and ground, or between both non-impulsed terminals of delta-connected windings and ground. When the neutral terminal is tested, the shunt is connected between line terminals and ground. Current shunt is usually made by a non-inductive resistor of 0.1~10 Ω; the value depends on whether a larger or small signal is required. A capacitor may be paralleled with the current shunt in order to limit the initial high frequency component due to the winding capacitance. The time constant of such compositive shunt shall be limited in 2 ns.

Impulse voltage is measured by the voltage divider which should be connected directly to the bushing terminal under test. The connection wire length between the divider and the bushing should be as short as possible. There are three types of

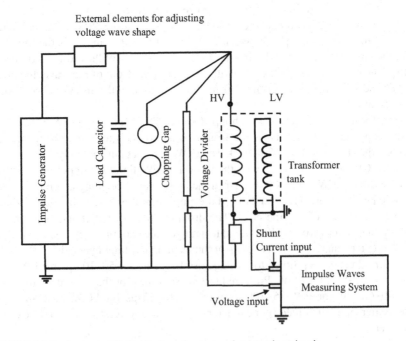

FIGURE 11.5 Layout of lightning impulse test and measuring circuit.

divider: resistive, capacitive and composited. The resistive divider may not be suitable for the switching impulse (SI) test due to the SI long tail wave. The composited divider is suitable for both lightning and switching impulse tests. The divider voltage ratio equals the impedance ratio of the primary side to the secondary side. A divider may have multi ratios by changing impedance of the secondary side, such as 200, 500, 1000, and 3000 to 1. The ratio used in test is selected such that the output voltage from the divider is in the range of input of recording instrument such as oscilloscope.

Number of chopping gap for the CW test depends on test voltage. The time to chop is controlled by a trigger. A resistor may be connected in series with the chopping gaps to limit overswing within 30%; when overswing is less than 30%, no resistor is required. The test set-up and measurement should be checked at 50% of test voltage before full voltage is applied. The chopping gap can consist of two square rods and works without a trigger.

A loading capacitor may be used as shown in Figure 11.5 to get required voltage waveform. The grounding of the tested transformer tank, voltage divider, chopping gap, loading capacitor and impulse generator should be made to one earthed copper bar. The number of stages of impulse generator used in the test is determined as

$$N = V_{BIL} / \left(\eta \cdot V_0 \cdot K_1 \right) \tag{11.5}$$

Where N is number of stages; V_{BIL} is test BIL level, kV; η is generator efficiency; V_0 is rated voltage per stage, kV; K_1 is voltage charging level per stage, %. For example, BIL level is 900 kV, CW level is 990 kV, assume generator efficiency $\eta = 90\%$, voltage per stage is $V_0 = 200$ kV, expect charging level is $K_1 = 80\%$, $N = 900/ (0.9 \times 200 \times 0.8) = 7$ (round up). If expect charging level is $K_1 = 70\%$, then $N = 8$.

The impulse test is conducted in sequence requested by standards or specified. Impulse waveshape is adjusted at RFW level, increasing or decreasing the front time by increasing or reducing front resistors' values in the generator; the tail time is adjusted by tail resistors' values in the generator in the same way. For low capacitance windings, the load capacitor is connected in parallel with the winding to obtain the required front time. For very low impedance windings, a special external circuit, such as a Glaninger Circuit [3], is used in the impulse test circuit as an additional option for better waveshape. Impulse generator multi-stage in parallel connection is also used to improve tail time.

A metal oxide varistor (MOV) connected in any part of the winding may cause a mismatch between FW and RFW. In such case it is advised to conduct two or more RFW tests at different voltage levels before and after the chopped wave, comparing the waveshapes of the same levels to see whether the mismatch is caused only by MOV, and not by insulation failure. The test sequence is usually RFW-RFW1-RFW2-FW-CW-CW-FW-FW-RFW2-RFW1-RFW.

Clearances from all high voltage leads, test devices and ground parts shall be checked. A rule of thumb for the lightning impulse test is 10 KV/25.4 mm. Unused impulse generator stages should be prevented from charging by removal of the charging resistor next to the charging stage; unused stages should be shorted and connected to the output of the last stage actually being used.

11.3.3 Failure Detection

Failure is detected by comparison between waveshapes of voltage and ground current in RFW and FW tests. The ground current waveshape is more sensitive to failure; voltage waveshapes should also be examined, although they are less sensitive. Judgment based on the ground current may not be suitable for chopped waves. Any difference in waveshapes may indicate failure. On the other hand, if the waveshapes are overlapped, the winding passes the lightning impulse test.

If deviations in the waveshape are observed, an investigation is needed to find the causes, and additional tests are performed when necessary. It should be noted that operations of protective devices, conditions of the external test circuit and external parts may also cause the deviation. If both voltage and current waves have distortions at the same time, the failure may be related to the transformer. If the distortion occurs only in voltage wave, check high voltage end setup first, or the voltage recording arm circuit. If the distortion occurs only in current wave, check the grounding current circuit connection, grounding resistor and shunt first. The use of the transfer function feature, incorporated in the recording systems, may help in the elimination of external circuit phenomena as possible causes of waveshape differences.

One way to detect failure is to listen to sound that is produced. Unusual sound within the tank at the moment of impulse could be an indication of failure. An experienced ear could identify the sound from between line end to tank wall, core side frame, core leg, between windings, and the winding top or bottom. Acoustic sensors are also used to locate failure. Further, opening manholes in the tank cover to see carbon smoking or bubbles immediately after impulse may help in locating failure. Measuring voltage, current, resistance, ratio and megger may also assist failure diagnosis.

11.3.4 Switching Impulse

The switching impulse test is conducted when unit voltage is 345 KV and higher, or specified [6]. Basic switching impulse level (BSL) is given in standards. The standard switching impulse waveshape is that the time to crest is at least 100 µs, the time to the first zero on the tail is at least 1000 µs and the time above 90% of voltage crest level is at least 200 µs. The tolerance of the voltage crest level is ±3%. Preferred sweep times are 150~250 µs for front time, 500 µs for voltage above 90%, and 1500~2500 µs for time to the first zero.

The test polarity is normally negative. The test is conducted to each line terminal bushing, one at a time. For a delta connection, one of two non-impulsed terminal bushings is grounded, another is left open; for a wye connection, neutral bushing is grounded, two non-impulsed bushings are jointed together and left open. In some cases, the voltage may have to be applied to LV winding to induce a voltage in HV winding; the voltage is measured from HV side.

Switching impulse builds up a magnetic field in the core due to its long tail and the open circuit of non-impulsed terminals. It may cause the core to saturate which may result in the time to first zero voltage less than 1000 µs, in addition, the reduced wave may be different in shape compared to the full wave. This effect is minimized or prevented by magnetically biasing the core in the opposite direction before applying the full wave. The magnetic bias is accomplished by applying several reduced impulses of opposite polarity before a full switching impulse. In some cases, biasing is unable to get 1000 µs for first zero voltage, and the shorter tail may be accepted because transformers in service see such reduced duration of switching surge. In the test, the reduce full wave should overlap the full wave until the core is saturated, a sequence of the impulses may be as −50%, +50%, +70%, −100%, +70%, −100%, +70%, −100%.

Tap connection is decided by manufacturer or specified, or at the tap which induces maximum voltage in non-impulsed terminals. Neutral terminals which are effectively grounded in service shall be solidly grounded or grounded through a low-impedance shunt during the switching test. CTs, PTs are connected in the same way as the lightning impulse test.

The voltage divider shall be either capacitive or composited because resistive type is unsuitable for the switching impulse test. HV bushings of tested transformer may be used as a divider to measure the voltage level. A load capacitor may be connected in parallel with the test terminal bushing to get the waveshape. Front and tail resistors in the generator which are generally larger than that used in lightning impulse test are adjusted to get the waveshape required, and time duration of voltage above 90%.

Switching impulse voltages are transferred between windings by turn ratios. However, higher voltages than calculated per turn ratio have been found between phase to phase in three-phase transformers, for example, the voltage between phase to phase of wye-connected windings is 1.5 times the applied voltage by turn ratio when the neutral is grounded. Tests found that voltage can be 1.85 times the applied voltage caused by heavy oscillation, so that the transformer insulation should be designed to take such oscillation. In the case that the oscillation is severe, and by the agreement of both manufacturer and user, the oscillations are damped by connecting non-impulse terminals to ground through resistors, usually connecting LV side non-impulse terminals to ground through resistors is easier. The oscillation should be checked at reduced full wave after the connection is made to ensure proper damping. A longer front time also helps to reduce the oscillations.

The switching impulse on the high voltage terminal results in an induced voltage on other windings; it can be lower or higher than its designed level. No additional test is required when the induced voltage on other windings is less than its designed level.

The failure is detected by superimposing voltage waveshapes of RFW and FW; it is not practical to pursuit total match. The waveshapes may differ slightly because of core saturation and test circuit. Grounding current oscillograms are generally not used for failure detection but can provide the information in troubleshooting. The unit passes the test when there is no sudden voltage collapse.

11.3.5 Transient Analysis

An instrument called a recurrent surge generator (RSG) is used to conduct transient analysis at low voltage such as 40 volts, with lightning full and chopped waveshapes. The test is performed on core and coil assembles in air before tanking, to study voltage distribution and insulation margins of windings. The test set-up is the same as the lightning impulse test; the measurement is taken from a point at winding to ground to see voltage to ground, or from point to point to see voltage differential, by means of thin metal rods with a shape tip to contact the winding conductors; the measured voltage waves is also recorded. RSG enables measurement of voltages between section to section, turn to turn, or tap to tap, phase to phase under FW or CW.

RSG is also used to study the stresses in windings under non-standard impulse waves such as fast front of switching surge, and to investigate failure cause. It offers useful information for winding insulation design.

11.4 APPLIED VOLTAGE

The applied voltage test, also called the Hipot test, is to verify insulations between windings and winding to ground. It is a power frequency withstand test using a single-phase source. The bushings of tested windings are connected together through a thin, bare wire, which then is connected to a high voltage terminal of the test power source. The bushing terminals may have to be shielded and connected by metallic tubes when partial discharge measurement is required during the test. Bushings of un-tested windings are connected together then to ground, and the transformer tank is grounded to the same ground of the testing power source.

A relief sphere gap, when it is in the test circuit, is set at a voltage 10% or more in excess of test value; the gap may be adjusted to produce a flashover at not more than 70% of test level before final gap setup. The gap can be also removed when it is confirmed that there is no risk of self-excitation [8].

The test voltage is measured by a calibrated voltage potential divider with an AC measuring device, the bushing in test may be calibrated and used as a voltage divider for the measurement. The test voltage is applied staring at or near zero, which is not higher than 25% of test level, then raising the voltage gradually to full level in no more than 15 seconds and avoiding overshoot. The voltage level is then maintained for 60 seconds, then reduced gradually in no more than 5 seconds to zero or minimum value before switching off the supply. The same procedures are repeated for each group of windings at their test levels.

Presence of smoke, bubbles in the oil, audible sound such as a thump or a sudden increase of current in test circuit may indicate failure. Each phenomenon should be investigated, and the test may be repeated or other tests are performed to assist in the failure investigation. In some cases, about 80% of the test level for one minute, called the "bubble-run" test, may be conducted. Then the unit rests for a few minutes before it is tested at full level; this pre-stress allows bubbles in oil to release.

11.5 INDUCED VOLTAGE AND PARTIAL DISCHARGE MEASUREMENT

11.5.1 INDUCED VOLTAGE TEST

This test is to verify the insulation integrities of windings and their line terminals to earth. It also tests the insulation integrity of turn to turn, section to section, and phase to phase. The tested transformer is connected the same way as service operation, the energized winding is usually the LV winding and it induces a voltage in HV winding. The single-phase unit is excited from a single-phase power source, and a three-phase unit from three-phase sources. The test level of HV winding is the target. The frequency used in the test is usually in the range of 130 to 400 Hz to prevent the core from saturation during the test [1].

The tap position should be such that the voltage level induced in HV windings is 1.5 times the maximum operation voltage, or follow related standards or specification. To avoid possible dangerous overvoltage due to self-excitation, wire jumpers may be used temporarily at HV bushing terminals to produce an air gap of about 25% of the bushing clearance to ground. When it is confirmed that there is no risk of self-excitation or "runaway", the wires are removed.

The neutral terminals and tank of the tested transformer shall be solidly grounded. Test duration of enhancement level in second is 7200 cycles divided by the generator frequency. It should not be less than 15 seconds regardless of the frequency.

Partial discharge is to verify both designed insulation integrity and production quality. When measurement of partial discharge (PD) or radio-influence voltage (RIV) is required, corona shields at high voltage bushing terminals may help in reducing external noise. For class I transformers, PD/RIV test duration may be extended to allow completion of the measurements on each terminal; the voltage is then reduced slowly to zero, or to one-hour level to continue PD measurement. For class II transformers, first, test voltage is raised to the one-hour level and held until PD/RIV readings of all terminals are taken. After verifying that there is no partial discharge problem, the voltage is raised to enhancement level and held for 7200 cycles; PD/RIV is measured when required. Then the voltage is reduced directly to one-hour level and kept for one hour, during which PD/RIV readings are taken at every 5 minutes on all required terminals. Finally, the voltage is slowly reduced to zero at the end of the one-hour period.

In some cases, a compensation circuit and an additional reactor may be used in the test circuit to reduce generator current or increase test capability.

Presence of PD/RIV levels in excess of limits, sudden current increase or trip off the test circuit, smoke or bubbles in the oil and audible sound such as a thump may indicate failure. An investigation should be carried out, and the test should be repeated or other tests are performed to verify whether failure has reoccurred. Oil samples may be taken for dissolved gas analysis (DGA) as information for the diagnosis.

11.5.2 PARTIAL DISCHARGE MEASUREMENT

Unlike the impulses and the hipot, the partial discharge test is non-destructive; it detects and locates the partial discharge inside the transformer which may be harmful to operation. The measured RIV in μV is on a quasi-peak [9] basis at a center frequency of 1 MHz in range of 0.85 MHz to 1.15 MHz. The measured apparent charge in pC is in the range of 50 to 400 kHz. Both test the partial discharge pulse but in different frequency. The test circuit connection is shown in Figure 11.6.

The test set-up uses the bushings of tested transformer as coupling capacitors, it is called the bushing-tap method. The input terminal of a measuring impedance unit is connected to the bushing test tap, the output terminal is connected to PD/RIV

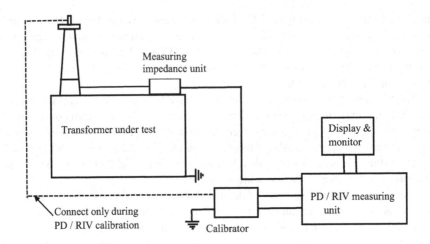

FIGURE 11.6   Partial discharge measurement circuit.

measuring system via a coaxial cable and the measuring impedance unit ground terminal should be connected to the transformer cover. A personal computer is connected to the measuring unit to monitor the testing. A calibration has to be conducted before the test; it establishes the relations of PD/RIV levels in the winding terminal and readings from the measuring unit. A calibrator is connected between bushing terminal H1 and ground, a calibration signal, say 300 Picocoulombs, is injected to H1, adjust the measuring unit to make the PD reading from measuring system the same as the injected level, or get a multiplier. The linearity of measuring system should also be checked in a range of values and the response of the measuring system. After bushing H1 is calibrated, the same calibration shall be carried out on other bushings one by one, the proportionality factor determined by the calibration shall be kept unchanged without adjusting any of the settings in the measuring system. The calibrated settings are not changed during the PD test. For RIV calibration, first, change the selector to RIV measuring mode, inject a certain amount of PD such as 100 μV to the bushing, adjust the measuring system to get reading equal or close to the injected value, or get a multiplier. Then the linearity response of the RIV system is checked by injections of different values. The calibration shall be carried out on each of test bushings; do not change any of the settings during the RIV test. Remove the calibrator before energizing the transformer.

When ambient PD/RIV levels are measured, they should not exceed half of the limit value. Record any random spikes, high ambient interference, etc., which could help in interpretation of the results, when the transformer is energized. As mentioned earlier, PD/RIV measurements are made at five-minute intervals on all required terminals during one-hour duration, a better way to test PD/RIV activity in transformer is to use multi-channel, such as a 6- to 8-channel measuring system to test all bushings simultaneously. The results are generally considered acceptable if the magnitude does not exceed 100 μV in RIV, 300 pC in PD or specified, and the increase in a one-hour period does not exceed 30 μV, 50 pC or specified, the PD/RIV levels during a one-hour period do not exhibit any steadily rising trend and there is no sudden and sustained increase during the last 20 minutes. Occasional high spikes should be disregarded. During initially increasing and finally decreasing the voltage, possible inception and extinction voltage may be recorded. The test may be extended or repeated to get an effective result.

Failure to meet the partial discharge acceptance criterion should lead to an investigation. A few things need to be checked out in the circuit, the oil level in the transformer, the pressure in the transformer tank, the lower pressure or vacuum in bushing head house, any ungrounded metal pieces on tank, the sharp point of any grounded metal piece facing bushing shields, whether the test circuit is properly set up and calibrated, any defective device/parts in test circuit, any loose connection, whether oil contains a high amount of moisture or particle, low oil dielectric breakdown voltage, insulation clearance and so on. All these factors may affect the test results.

By comparing measured PD patterns to typical PD patterns (PD source statistical behavior), it may give indications of the PD source and its possible location, such as a metal tip in oil or in solid insulation material, a tip electrode on the surface of solid insulating material, electrically conductive particles in oil or in solid insulation, bad metal contact, bubbles in oil or in solid insulation, bubbles due to the humidity of the insulating system, external noise, ground side shape points, cavity contact directly with metal electrode, or inside insulation.

Dangerous PD sources are listed as following, PD inception or extinction voltage is below the rate voltage, the source is in solid insulation, PD continuously changes in pattern, the amplitude increases with the time during the test and the number of signals per cycle increases with time during the test.

High PD level may decrease after the unit is continuously energized for an extended time, or after oil is reprocessed and the unit rests a longer time before being retested. Increasing the pressure inside the tank can also reduce PD level, so the range of the pressure should get a permission.

An acoustic wave is produced and emitted by the PD source, the measurement of time delay between PD electric signals and acoustic signals may locate the PD source [10]. Using the PD electric signal as an oscilloscope trigger, and detecting the acoustic signals from at least three acoustic sensors placed at the transformer tank wall, the location of the PD source is calculated from the time delay between the electric and the acoustic signals by the triangulation method. The velocity of acoustic waves in oil is about 1400 m/s. Transformer having a complicated insulation structure could affect the acoustics wave propagation. PD sources hidden in solid insulation are not easily detected due to the different propagation velocities in different materials and the reflection of waves. The PD source in the main insulation between coils is difficult to be located due to the insulation barriers and outer winding. For a single PD source mode, an acoustic PD locating test should be performed at a voltage level a little higher than that of the PD inception voltage.

To help in investigating the PD source, spectrum analyzer, advanced PD-system and acoustic locating may work in tandem to get more useful information. PD tested under different conditions also can give more required information for locating. The final step may be internal inspections, or untanking the core and coil assembly to find defective parts.

11.6 LOAD LOSSES AND IMPEDANCE

11.6.1 LOAD LOSSES AND POSITIVE IMPEDANCE

The loss produced when load current flowing through windings is called load losses. It includes I^2R and eddy current losses of the winding conductor, and stray losses from metallic structural parts. Load losses are generally tested by a short-circuit method [1]. The voltage required to circulate rated current through a winding while another winding is short-circuited is call impedance voltage. Impedance voltage is usually expressed in percentage of the rated voltage of winding stressed.

Four wires of the three-phase measuring system is preferred to test three-phase transformers. The impedance voltage is measured directly at the bushing terminals of the transformer being tested. Measurements are made at rated current and frequency at nominal taps and extreme taps or specified, positive impedance in percent $Z\%$ is

$$Z\% = \frac{V_t \cdot I_r}{I_t \cdot V_r} \times 100 \qquad (11.6)$$

where V_t and I_t are tested voltage and current respectively, and V_r and I_r are rate voltage and current respectively. Load losses are measured as a whole; the stray loss is obtained by subtracting I^2R losses from the load loss. The temperature requirement is same as cold resistance and no-load loss testing; the difference in winding temperature shall not exceed 5°C before and after the test. The frequency shall be within ±0.5% of the rated. The metering maximum correction is limited to ±5% of the measured losses. The transformer is energized from the HV side, the LV winding is short-circuited by a copper bus bar which should have a cross-sectional area no less than the corresponding transformer leads; 1000 amps per 650 mm² is used as a reference. The test set-up is shown in Figure 11.7.

When LV bushings of a large transformer are short-circuited, it represents a highly inductive load, its power factor is on the order of 2 or 3 percent, and a capacitor bank is needed to produce capacitive compensating current to reduce the load on the power

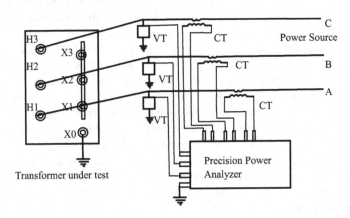

FIGURE 11.7   Load loss test set-up.

supply. The capacitor bank may be connected to a test loader (testing transformer) HV side or LV side. The amount of capacitive compensation is based on the tested transformer. Careful selection of the compensation is required to avoid any risk of voltage and current running away. Internal CTs, PTs and core ground connections shall be the same as no-load loss testing. In some cases, the test voltage is applied to the LV side and HV bushings are shorted.

The test voltage is gradually increased until the current reaching rated value, the voltage, ampere and Watt shall be recorded simultaneously. The measuring shall be made in a short period of time to minimize heating effect. The first test should be made at the winding connection and taps which the loss and impedance are guaranteed. A correction of the test system phase angle on tested loss should be made, tested load losses and impedance should also be corrected to reference temperature, which is rated winding temperature rise plus 20°C. The tested load losses for temperature rise test should also be corrected to the appropriate reference temperature.

When not specified by user, the tolerance of tested impedance away from guaranteed for a two-winding transformer should be ±7.5% if guaranteed impedance is greater than 2.5%, ±10% if the guaranteed impedance is 2.5% or less; for three windings and zigzag winding transformers, the tolerance is ±10%. Total loss that is no-load loss plus load loss shall not exceed guaranteed loss by more than 6%.

When testing three winding transformers, the average rms values of three line currents are adjusted to desired values, and if the three line currents are noticeably unbalanced, the tested losses will be higher than they should be. When the voltage is measured at a a point before the bushings, it can also cause higher losses due to the voltage drop by the current flowing through the leads. The test connection pairs for three winding transformers may be HV-LVX, HV-LVY and LVX- LVY, the pair of HV-(LVX + LVY) may also be carried out.

When there is trouble in the load loss and impedance test, the single-phase power used to test the three-phase transformer may give more information for troubleshooting.

Impedance percentage can be expressed in a plural as $Z = r + jx$, where r is the real part of impedance related to the load loss, x is the imagined part of impedance related to the reactance. On the other hand, with the plural format of impedance, the load losses can be estimated. For example, percentage of impedance $Z = 0.285 + j12.01$ at 30 MVA, the load losses = $(0.285/100) \times 30,000,000 = 85,500$ (W).

11.6.2 Zero Sequence Impedance

Zero sequence impedance is the impedance to the simultaneous current flow in all three phases when a single phase voltage is applied between one set of line terminals connected together and neutral. Zero sequence impedance can be measured only when a physical neutral is brought out [1]; with more than one neutral, multiple tests are needed to get each branch value of the zero sequence impedance network. For some units, zero sequence impedance is nonlinear and dependent on the value of test current, so more than one set of measurements should be taken to characterize its nonlinear behavior.

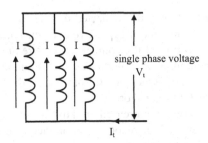

FIGURE 11.8 The connection for zero sequence impedance test.

When a transformer has a delta winding, the applied voltage should be such level that the current in the delta winding should not exceed the rated value. For a unit without delta winding, the applied voltage should not exceed 30% of the rated line to neutral voltage of the winding under test, nor should the phase current exceed its rated value. Based on the test voltage and current, zero sequence impedance is

$$Z_0\% = 300 \times \frac{V_t}{I_t} \cdot \frac{I_r}{V_r} \tag{11.7}$$

Where V_t is test voltage, I_t is test total input current, V_r is rated phase to neutral voltage and I_r is rated phase current. It needs to verify that rating of neutral bushing and internal neutral connections are adequate for test currents; this information can be provided from the design. All delta windings should be closed.

For wye–delta connected units, apply single phase voltage between shorted three-line terminals of wye windings and neutral. The test current by applied voltage should not exceed the rated phase current of delta winding.

For wye–wye connected unit, four measurements are taken. First, apply voltage between $H_1H_2H_3$ and H_0; open X_1, X_2, X_3, X_0. Second, apply voltage between $H_1H_2H_3$ and H_0. Shorten X_1, X_2, X_3, X_0. Third, apply voltage between $X_1X_2X_3$ and X_0, Open H_1, H_2, H_3, H_0. Last, apply voltage between $X_1X_2X_3$ and X_0, and shorten H_1, H_2, H_3, H_0. Only three test results are needed to get the values of a zero sequence impedance network.

For wye–wye–wye connected units, four measurements are taken. First, apply voltage between $H_1H_2H_3$ and H_0, shorten X_1, X_2, X_3, X_0, open Y_1, Y_2, Y_3, Y_0. Second, apply voltage between $H_1H_2H_3$ and H_0, open X_1, X_2, X_3, X_0, open Y_1, Y_2, Y_3, Y_0. Third, apply voltage between $X_1X_2X_3$ and X_0, open H_1, H_2, H_3, H_0, open Y_1, Y_2, Y_3, Y_0. Last, apply voltage between $Y_1Y_2Y_3$ and Y_0, open H_1, H_2, H_3, H_0, open X_1, X_2, X_3, X_0. From these test results, the zero sequence impedance network can be achieved.

Zigzag unit is often used as a grounding transformer; its zero sequence impedance is tested in the same way discussed earlier. If the grounding transformer is short time rated, test current and duration may need to be specified.

11.7 TEMPERATURE RISE

The temperature rise test, also called heat run test, is one of design tests. It measures winding and oil temperature rises caused by total losses. The test should be conducted at combination of connection and taps that give the highest winding and oil temperature rises [1]; it generally is the tap of the highest losses. The highest losses usually happen at the highest current tap where total turns of tap sections or whole tap winding are not included; it means that these turns are not be tested, and any possible gassing sources from these turns could be missed. To test all winding turns, the test should be carried out at maximum voltage tap with the highest losses the transformer could generate, the gradients under test current should be corrected to the highest rated current, in such a way that the highest oil temperature rises are obtained, and the test current flows through all winding turns to check out any possible issue.

Thermocouples, resistance temperature detectors (RTD) or suitable temperature sensors are generally used in the test. Top oil temperature is directly measured at location of 50 mm below the oil surface in the tank, the average oil temperature should be the top oil temperature minus half of the temperature drop between top oil and bottom oil in the radiator or cooler. It also can be the drop between the surface temperatures of the inlet and the outlet of the radiator when oil temperature in the radiator cannot be measured directly. The ambient air temperature should be measured using three thermocouples spaced uniformly around tested transformer by 1 to 2 meters away and at locations about half of the tank height. Fiber optics are sometimes used to measure winding temperatures directly in recent years.

Total losses used are the sum of no-load losses and load losses tested and converted to ones at the temperature equal to rated winding rise plus 20°C. The short-circuit test method is usually used, the transformer is energized as load losses test connection by shorting one or two sets of windings and to make sufficient current flow through the windings to produce required total losses. The applied voltage, current, Watts, all temperature readings of top oil, inlets and outlets of radiator, and air ambient should be recorded in a time interval period, such as 30 minutes. The time when the fans or pumps are turned on should be recorded, too. Liquid temperature gauge, winding temperature gauge at the transformer tank may be pre-calibrated and used to record the temperature during the test.

When top oil temperature rise does not vary more than 2.5% or 1°C, whichever is greater, during a consecutive three-hour period, the top oil rise reaches the steady state condition. Afterward, the test current is reduced to rated current and maintained for one hour before shutdown is performed; it is called full current run. The shutdown starts at the moment that the applied power is cut off, and it marks as time zero of the cooling curve. Cut off the power of cooling fans if there is any, disconnect the short-circuit bus bar on LV bushings, set up the resistance measuring circuit that should be same as one for cold winding resistances measurement, and take a reading of winding hot resistance at each time interval, say 15 seconds. The first reading should be obtained no longer than 4 minutes after shutdown; in such way, a series of hot winding resistance with time in a 10~15 minute period is collected. The resistance-time data should be fitted to an exponential decay curve by the method of least square.

This curve, called a cooling curve, is used to determine the hot winding resistance at the instant of shutdown. The cooling curve shall be made on all terminal pairs at maximum nameplate ratings by repeating full current run and shutdown afterwards. The terminal pair that has the highest winding temperature should be retested only for the hot resistance measuring (shutdown) at base rating or at overload test.

HV and LV winding resistances can be measured in a series connection manner. In some cases two independent DC sources are used for different windings testing, one for HV and another for LV.

The time to perform the temperature test can be shortened by applying overload or reducing the cooling capacity at the initial stage. The top oil temperature increase rate should not exceed 10°C/hour. The test should be returned to required total losses and rated cooling conditions when the top oil temperature reaches the predicted value, then the test continues till the top oil temperature reaches its steady state condition. On the other hand, pumps are left running at all times for pumped units.

The cooling curve achieved from the tested data is extrapolated back to time zero to get the hot resistance R_{t0} at the instant of shutdown. The copper winding temperature at the instant of shutdown T_0 is calculated by

$$T_0 = \frac{R_{t0}}{R_{co}} \times \left(T_{co} + 234.5\right) - 234.5 \tag{11.8}$$

where R_{co} is cold resistance at temperature T_{co}. Ambient temperature $T_a = (T_{1a} + T_{2a} + T_{3a})/3$, here T_{1a}, T_{2a} and T_{3a} are ambient readings should be taken at the instant of shutdown. Top oil rise is $TOR = T_p - T_a$, where T_p is measured top oil temperature mentioned earlier. Average oil rise is $AOR = TOR - \Delta T$, where $\Delta T = (T_{tr} - T_{br})/2$. Here T_{tr} and T_{br} are average readings of top and bottom temperatures of radiators respectively. Winding gradient to oil is $\Delta T_g = T_0 - AOR - T_a$. Winding temperature rise is $T_w = AOR + \Delta T_g$. Winding hot spot rise is $HSR = TOR + k\,\Delta T_g$, here k is hot spot factor, in range of 1.1 to 1.3 in general.

If the tested current I_t is different from rate current I_r, winding gradient ΔT_g has to be corrected to rated current by factor $K_i = (I_r/I_t)^{2m}$. Here, $m = 0.8$ for ONAN, ONAF, OFAF and OFWF, $m = 1.0$ for ODAF and ODWF. Due to limitations of test equipment, the test current no less than 85% of rated current may be used.

If injected losses W_t is different from required losses W_r, the liquid temperature rise has to be corrected by factor $K_w = (W_t/W_r)^n$. Here $n = 0.8$ for ONAN, 0.9 for ONAF, OFAF and OFWF, 1.0 for ODAF and ODWF. The correction is available for test loss being within 20% of the required loss.

No altitude correction is needed when the tested is conducted at altitude of 1000 m or lower. When a transformer is tested at altitude less than 1000 m but operates at an altitude above 1000 m, temperature rises need to be corrected by standards or specified.

For an overload temperature test, pre-overload conditions, overload levels and duration shall be specified. The hot resistance measurements should be made at the hottest phase found from the previous test. An example of the test sequence is showed in Figure 11.9. Oil may be sampled before and after each test stage for

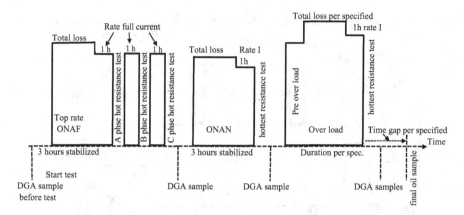

FIGURE 11.9 Temperature rise test sequence.

DGA analysis. Thermal scan or thermal photography may be performed at total losses run before full current run to check temperatures of tank wall and cover, and the radiators and bushings.

11.8 AUDIBLE SOUND LEVEL

Transformers produce unwanted sound or noise during operations, and standards and regulations exist to limit the sound below certain levels. The sounds produced by a transformer are from its core, windings and cooling device. The sound from the core and running fans are called no-load sound; the sound produced by the load current in windings is called load sound. Its magnitude is decided by the loading level, which is a significant contributor to the total sound in some units.

Test setup of no-load sound is same as no-load loss test setup, The transformer is energized by rated voltage and frequency at the nominal tap position [1,8], or the position having the highest sound, or specified. For variable flux transformers, the test should be conducted at the highest flux density tap position. Test setup of load sound is same as the load loss test setup, transformer is energized to rated current and frequency at nominal tap position, or the position having the highest current, or specified. Load sound level may be measured at top rating at 2 meters measuring contour as no-load sound measurement. The total sound level is the combination of no-load sound and load sound levels.

The transformer being tested is placed asymmetrically in test area, a minimum of 3 meters away from acoustical reflecting surface. The transformer should sit directly on the ground floor. A reference sound-producing surface is found by a taut string tied around the periphery of the transformer including radiators/coolers, terminal chambers, tap changers and tubes, etc. It may not include bushings and minor extensions such as oil gauges, valves or projections at or above cover height. Three filtered sound levels, such as A-weighted, C-weighted and one-third octave band, are usually used in the test, the sound level meter is calibrated before conducting the test.

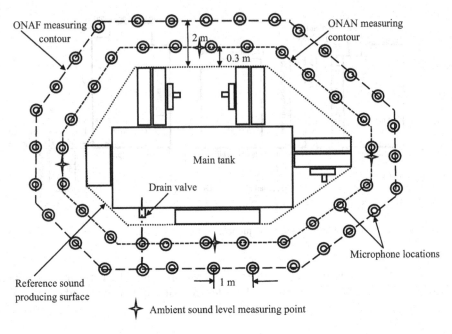

FIGURE 11.10 Microphone measuring position.

Microphone measuring points, as shown in Figure 11.10, shall be on a contour 0.3 meter away from the reference sound producing surface at intervals 1 m in a horizontal direction for an ONAN rating for which fans are not running, 2 meters away for an ONAF rating for which fans are running.

The first measuring point should be aligned with main drain valve or specified, then proceeding with the measurement clockwise along the measuring contour. The microphone heights should be 1/3 and 2/3 of tank height when it is 2.4 meters or more, half of height when it is less than 2.4 meters. Four points as minumum shall be measured such as on a small transformer.

The sound levels measured when a transformer is energized are actually combined sound levels of the transformer and ambient background. To get the sound level of the transformer itself, the ambient background level shall be measured at minimum of four locations around the de-energized transformer at the center of each tank wall before and after measurements of energized transformer sound levels. These measurements are corrected as such, if the difference between ambient background and transformer sound level is 5, 6, 7, 8, 9 and 10 dB, the deduction is −1.6, −1.3, −1.0, −0.8, −0.6, and −0.4 respectively [1]. No correction is needed when the difference is over 10 dB, and if the difference is less than 5 dB, a correction of −1.6 dB is suggested in the case that it is desired to know whether the sound level is over the limit. After the ambient correction, the average sound pressure level is computed. Afterwards, the wall sound reflection correction can be conducted per standards.

The combination level of no-load and load sound is obtained per equations in Chapter 9. The combined sound level goes up to 3 dB maximum higher than higher one of no-load and load sounds.

The sound level at different contours is achieved by adding or subtracting the value of $10 \times lg(S_1/S_2)$, here S_1 and S_2 are the length of different contours respectively. The closer the contour is to the transformer, the higher the sound levels.

Based on measured load sound level of rating MVA_1, the sound level of rating MVA_2 at the same measuring contour is calculated as $L_2 = L_1 + 40 \times lg_{10}(MVA_2/MVA_1)$. Here, L_1 is the load sound measured at MVA_1.

11.9 OTHERS

11.9.1 OIL

An oil dielectric breakdown test should be conducted in accordance with related standards; the oil is usually sampled from the drain valve. The oil background conditions and the time of oil sampled should be noted its the reports. The breakdown voltage of good-quality oil per D1816 [11] is higher than 30 kV and 50 kV in 1 mm and 2 mm electrode gaps respectively. High voltage units need high oil breakdown voltage. Purification of oil can help a unit pass an oil dielectric test, but such oil may not reflect the unit service condition.

An oil test may include measurements of moisture content, particle count, dielectric dissipation factor, interfacial tension IFT, acid neutralization number, etc. The moisture content is required less than 5 ppm before testing a new transformer.

11.9.2 DISSOLVED GAS ANALYSIS (DGA) [12]

The decompositions of oil and cellulose insulation could occur in a transformer due to abnormal thermal and electrical faults such as partial discharge and arcing. Carbon oxides (CO, CO_2) and some hydrocarbons (H_2, CH_4, C_2H_2, C_2H_4, C_2H_6, etc.) could be generated and dissolve in the oil. The amount and types of these byproducts are related to the temperature produced from the faults. Analysis of different gases by a gas chromatograph helps to know the characteristics of faults in testing or service. Generally, the faults are divided into two basic types, thermal and electrical. The thermal faults are classified into three ranges of temperatures: T < 300°C, 300°C < T < 700°C and T > 700°C. The electrical faults are classified as low energy discharge and high energy discharge or Arcing. High energy discharge can produce a temperature from 700°C to 1800°C, with acetylene C_2H_2 as the key gas. An initial electrical discharge could develop into a thermal problem. Careful analysis on DGA result could find out the true cause of fault. Oil sampling and DGA may be repeated to monitor trend of gases increasing with energizing time, and it also gives further information on the assessment of the transformer's condition. It is important to know if a fault that generates gases is active.

Oil syringes are generally used to collect oil samples; the volume of the syringe is commonly in 50 ml range. The cleanliness of the syringe is important, and the collection of oil sample should contain no external air bubbles or gases.

11.9.3 Short-Circuit

Short-circuit tests are to verify mechanical withstand capability in short circuit situation. The tests are expensive and are unable to be performed in most of manufacturer's test laboratories; they are conducted in special laboratories.

Routine tests except a lightning impulse test should be carried out prior to a short-circuit test on tap position that produces the most severe mechanical stresses per calculations. The temperature of top oil at beginning of the short-circuit test should be between 0°C and 40°C. A pressure relief device and gas relay should be mounted.

The test is carried out by one of the two methods. The post-set method involves closing a breaker at faulted terminals to apply a short-circuit to ground to energized transformer. The pre-set method involves closing a breaker at source terminals to apply energy to short-circuited transformer. The post-set method is closer to the practical condition of the faults and should be used in the test.

The test should include measurements of symmetrical and asymmetrical currents. A synchronous switch or another controlled switching device is adjusted to obtain asymmetrical current wave in the phase under testing. To establish the required source voltage or closing switch time, calibration tests should be carried out at 50% of specified current. A test current of 95% or more of the specified current should be counted. When the post-set method is used, the voltage should not exceed 110% of the rated voltage or specified.

Tests are conducted six shots per phase, four under symmetrical and two under asymmetrical currents. The duration of the shots should follow standards or specified. For example, the duration is 0.25 second for the asymmetrical current test, and 0.5 or 1 second for the symmetrical current test. Oscillography recordings are taken for the voltage and current during the tests.

The following conditions need to be met for satisfactory performance. First, no abnormal phenomenon is observed during the test. Second, no pressure relief device or gas relay is activated. Third, no abrupt change or anomaly is recorded on test voltage and current wave shapes. Fourth, there is no indication of any change in physical condition of core and coils by visual inspection. Fifth, less than 2% change on impedance is measured before and after the short-circuit test. Sixth, less than 5% increase of no-load current comparison is measured before the short-circuit test. Seventh, the transformer withstands the standard dielectric tests following short-circuit tests. Eighth, there are no abnormalities in dissolved gas analysis.

11.9.4 Diagnostic Tests

11.9.4.1 Sweep Frequency Response Analysis (SFRA)

This test is to find any movement of the windings and core after unit is moved from place to place. From an electric circuit point of view, a transformer is a complicated network of resistance, inductance and capacitance, or RLC circuit. It behaves like an RLC filter and outputs different voltages at different frequencies depending upon the nature of RLC circuit. A particular pattern can be obtained by plotting the output voltages against the corresponding frequencies for a particular winding. If the same plotting is conducted before and after events of transportation or short-circuit faults, and superimposed to observe deviation between the two graphs, whether physical displacement and deformation occurred in the winding can be assessed. The comparison also can be carried out between phases in the same tap positions, or between sister units.

A SFRA curve is used as a benchmark or fingerprint, produced by manufacturers before shipment. SFRA test is carried out the curve by injecting various discreet frequencies in the range of 10 Hz to 20 MHz.

Depending on response of core and winding, an SFRA test is carried out with an open circuit connection. The tests are performed on a winding with all other bushings floating. For isolation of winding response, the SFRA test can also be done with a short-circuit connection; the test is performed on HV bushing by shorting all LV bushings together, and the neutral bushing is not joined to the shorting process and left floating. These tests will help to narrow down the areas where the problem might be. The tap position of the transformer should be that all turns are included. For transformers having an on-load tap changer (OLTC), one more test may be carried out with OLTC out of circuit. The test connections shall follow related standards or specified.

Mismatch between SFRA curves before and after transport or short-circuit faults may indicate core or coil movements. To characterize the movement, SFRA curves could be divided into four regions in frequency band. Region 1 (less than 2 kHz) may indicate open circuits, shorted turns, core deformation and residual magnetism. Region 2 (2 kHz to 20 kHz) may indicate bulk winding movement between windings and clamping structure. Region 3 (20 kHz to 400 kHz) may indicate deformation within the main or tap windings. Region 4 (400 kHz to 1 MHz) may indicate movement of the main and tap windings and ground impedance variations. These are just a general rule of thumb; the degree of mismatch depends also on transformer design and test conditions. Testing above 2 MHz tends to be dominated by variations in grounding practices for the test leads.

11.9.4.2 Dielectric Frequency Response (DFR)

The DFR test is similar to the power factor (PF) or dissipation factor test. PF is tested at power frequency, while DFR tests PF at multiple frequencies, usually in the range of 1000 to 0.001 Hz. The DFR test provides much more information on dielectric properties of the insulation system of the transformer. A variable frequency voltage is applied to the transformer, and a complex calculation is performed to digitize and process voltage and current signals.

DFR provides evaluation of moisture content which relies on a database of dielectric response measurements of oil-impregnated pressboard samples with known moisture levels and temperature. The test data is compared to the response of the modelled insulation system. The moisture in cellulose and oil conductivity values are then optimized for a best fit of the calculated response curve.

In a DFR curve, in lower and higher frequency ranges, the curve shape is determined by solid insulation moisture content and insulation geometry, over a mid of range is a simple straight line which is determined by oil conductivity. At lower temperature, the response curve shifts to the left, and testing to lower frequencies may be required.

DFR is good for troubleshooting on power factor problems in manufacturing or field service. A transformer with a high power factor may be caused by (1) moisture in solid insulation; (2) high oil conductivity due to aging, overheating, moisture or pollution; (3) carbon tracking in cellulose; (4) chemical contamination of solid insulation; (5) high electrical resistance in magnetic core circuit. Identifying the causes may be accomplished by comparing the DFR to other DFR functions from transformers with known defects or to results from laboratory tests.

REFERENCES

1. *IEEE Std C57.12.90-2015, IEEE Standard Test Code for Liquid-Immersed Distribution, Power, and Regulating Transformers*, 2015.
2. Ake Carlson, et al, *Testing of Power Transformers, Routine Tests, Type Tests and Special tests*, 1st edition, Pro Print GmbH Düsseldorf, Zurich, Switzerland, 2003.
3. *72A-1230 Rev F, M4000 Insulation Analyzer User Guide,* Doble Engineering company, Watertown, Massachusetts, 2000.
4. Ake Carlson, et al, *Testing of Power Transformers and Shunt Reactors, Routine, Type and Special Tests*, 2nd edition, Pro Print GmbH Düsseldorf, Zurich, Switzerland, 2010.
5. *IEEE Std C57.98-2011, IEEE Guide for Transformer Impulse Tests*, 2011.
6. *IEEE Std C57.12.00-2015, IEEE Standard for General Requirements for Liquid-Immersed Distribution, Power, and Regulating Transformers*, 2015.
7. *Haefely Recurrent Surge Generator RSG 482 User Manual*, 2011.
8. ABB Management Services Ltd Transformers, *Service Handbook for Transformers*, 3rd edition, Zurich, Switzerland, 2007.
9. IEEE Std C57.113-2010, *IEEE Recommended Practice for Partial Discharge Measurement in Liquid-Filled Power Transformers and Shunt Reactors*, 2010.
10. IEEE Std C57.127-2007, *IEEE Guide for the Detection and Location of Acoustic Emissions from Partial Discharges in Oil-Immersed Power Transformers and Reactors*, 2007.
11. ASTM, D1816-12, *Standard Test Method for Dielectric Breakdown Voltage of Insulating Liquids Using VDE Electrodes*, 2019.
12. ASTM, D3612-02, *Standard Test Method for Analysis of Gases Dissolved in Electrical Insulating Oil by Gas Chromatography*, 2017.
13. IEEE Std C57.149-2012, *IEEE Guide for the Application and Interpretation of Frequency Response Analysis for Oil-Immersed Transformers*, 2012.
14. IEEE Std C57.161-2018, *IEEE Guide for Dielectric Frequency Response Test*, 2018.

Index

Printed in the United States
By Bookmasters